A Journey Through The Universe

A Journey Through
The Universe

*A traveller's guide from the centre of the sun to the
edge of the unknown*

NEW SCIENTIST

First published in Great Britain by John Murray Learning in 2018
An imprint of John Murray Press
A division of Hodder & Stoughton Ltd,
An Hachette UK company

This paperback edition published in 2022

3

A CIP catalogue record for this title is available from the British Library

B format ISBN 978 1 529 38197 9
eBook ISBN 978 1 473 62985 1

Typeset by KnowledgeWorks Global Ltd.

Printed and bound in Great Britain by Clays Ltd, Elcograf S.p.A.

John Murray Press policy is to use papers that are natural, renewable and recyclable
products and made from wood grown in sustainable forests. The logging and
manufacturing processes are expected to conform to the environmental regulations of
the country of origin.

John Murray Press
Carmelite House
50 Victoria Embankment
London EC4Y 0DZ

Nicholas Brealey Publishing
Hachette Book Group
Market Place, Center 53, State Street
Boston, MA 02109, USA

instantexpert.johnmurraylearning.com

Contents

Series introduction

New Scientist's Instant Expert books shine light on the subjects that we all wish we knew more about: topics that challenge, engage enquiring minds and open up a deeper understanding of the world around us. *Instant Expert* books are definitive and accessible entry points for curious readers who want to know how things work and why. Look out for the other titles in the series:

Contributors

Editor: Stephen Battersby is a physics writer and consultant for *New Scientist*

Series Editor: Alison George

Instant Expert Editor: Jeremy Webb

Writers: Jacob Aron, Anil Ananthaswamy, Stephen Battersby, Emily Benson, Rebecca Boyle, Marcus Chown, Stuart Clark, Andy Coghlan, Rachel Courtland, Leah Crane, Ken Croswell, Sarah Cruddas, Pedro Ferreira, Will Gater, Conor Gearin, Lisa Grossmann, Adam Hadhazy, Alice Hazelton, Nigel Henbest, Hal Hodson, Rowan Hooper, Adam Mann, Dana Mackenzie, Maggie McKee, Mika McKinnon, Hazel Muir, Sean O'Neill, Shannon Palus, Aviva Rutkin, Govert Schilling, Sarah Scoles, David Shiga, Michael Slezak, Joshua Sokol, Colin Stuart, Richard Webb, Chelsea Whyte, Sam Wong, Aylin Woodward.

Introduction

This is a journey into space. On board a spacecraft of advanced design we will visit the highlights of the known universe – swooping among the hundreds of quirky worlds in our own solar system, then out into the Milky Way to meet seething unstable stars and exotic exoplanets, through the mysteries of the intergalactic void, to far-flung galaxies where giant black holes shine, and beyond to the exploding stars and collisions that create some of the most distant beacons we can see. For convenience, our embarkation point will be nearby, a mere 499 light seconds away, at an astronomical object that has been observed ever since eyes evolved.

Stephen Battersby, Editor

I
Star one

An unremarkable star in cosmic terms, the sun dominates our solar system, our sky and our lives. Within its core, protons fuse together to form helium nuclei, generating the heat that warms Earth along with a horde of ghostly neutrinos. As well as giving us light and life, the sun has the power to bring chaos to civilization – unless we can better understand its magnetic mysteries.

The strangest star

Billions of stars fill our galaxy. Many burn bright, destined to become supernovae, while others are dim burnouts. They come alone and in pairs; with or without planetary companions. We have searched the far reaches of the universe in the hope of understanding the stars... but ultimately everything we know is based on our local reference point, the sun.

It is made of **plasma** – ionized gas. It fuses hydrogen in its core. It blasts us with radiation and life-giving light. As stars go,

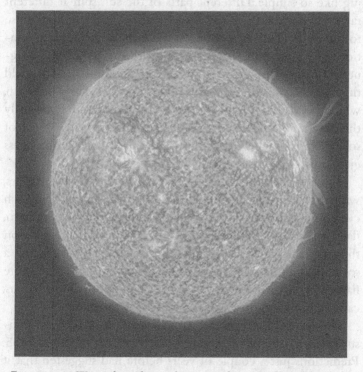

FIGURE 1.1 We tend to take our home star for granted but it is a truly remarkable object, with its own rain, tornadoes and plasma jets.

it is roughly middle-aged, having been around for 4.6 billion years. And it probably has 5 or so billion more to go before it swells into a **red giant** that consumes Mercury, Venus and Earth. Yet our home star remains mysterious, abounding with strange phenomena...

A magnetic calendar

Our planet takes 24 hours to spin once on its axis and 365 days to travel around the sun. The sun's own schedule is nothing like so simple. Different parts of the sun spin at different rates. So while a day at the equator lasts 25 days, regions close to the poles take a few days longer to make a complete rotation. This uneven spin leads to distortion in the sun's magnetic field. As the equator spins, it drags the magnetic field that connects to the sun's pole. The sun's field gets wound up, which builds tension, like twisting a rubber band. Eventually the magnetic field snaps and releases energy in the form of solar flares or huge projectiles of plasma called coronal mass ejections (CMEs).

This activity follows a cycle that lasts roughly 11 Earth years, with the overall magnetic field reversing its direction in each cycle, giving the sun its own calendar. During a solar minimum, flares are few, and so are sunspots (dark patches that appear on the sun's surface, marking intense magnetic fields). During a solar maximum, more sunspots burst out, and there are more flares and CMEs. Sometimes CMEs can hit Earth and affect us, causing blackouts on Earth and damaging satellites.

The last solar maximum, around 2012 to 2015, was unusually calm, one of the weakest since records began in 1755. Predictions just a couple of years before had suggested that it would be a scorcher, which shows how little we understand solar cycles.

Blowing bubbles

As the sun follows its 11-year cycle, it alters its output of solar wind, X-rays, ultraviolet and visible light.

The sun provides nearly all the energy that drives Earth's climate – 2500 times as much as all other sources combined. Solar activity has been partly responsible for warm and cool periods in the past, and today low solar activity contributes to cold winters in northern Europe and the US, and mild winters over southern Europe – although this is a small effect compared with global warming.

We now understand what's going on a little better thanks to a space-borne instrument called TIM, launched by NASA in 2003. TIM keeps tabs on the spectrum of radiation the sun emits, and detects subtle changes in energy output so scientists can distinguish between human causes of climate change and purely natural causes we can't control.

Changes in the sun's output affect much more than just our climate, however. During a solar minimum, the stream of charged particles called the solar wind flows from the poles at a much higher speed, so there's more pressure pushing against material from interstellar space. This increases the size of the heliosphere, a huge magnetic bubble of charged particles that the sun blows around itself, stretching way out beyond Pluto. During a solar maximum, the sun's magnetic fields are more knotted up and not as much wind escapes, so the heliosphere contracts.

Rain on the sun

We know the sun affects weather on Earth and in space, but it has its own dramatic weather. Superhot plasma that surrounds the sun forms the solar corona. Some of this plasma streams away to form the solar wind, but it can also rain back to the surface.

Though this coronal rain was predicted about 40 years ago, we couldn't see or study it until our telescopes became powerful enough. It works a bit like the water cycle on Earth – where vapour warms, rises, forms clouds, cools enough to condense into a liquid and falls back to the ground as precipitation. The difference is that the plasma doesn't change from gas to liquid, it simply cools enough to fall back down to the solar surface.

This all happens very quickly and on a gargantuan scale, with droplets the size of countries plunging from heights of 63,000 kilometres – about one-sixth of the distance from Earth to the moon.

The sun has tornadoes too. Swirling solar plasma creates a vortex, which causes magnetic fields to twist and spiral around into a super-tornado that reaches from the surface into the upper atmosphere.

Defying thermodynamics

Solar tornadoes are bizarre enough on their own, but they might help explain one of the sun's weirdest characteristics: its atmosphere is hotter than its surface. At just 5700 kelvin, the sun's surface is frigid compared with the corona above, which can reach temperatures of several million kelvin.

Generally, an object cools as it moves away from a heat source. A marshmallow will toast faster when it's closer to a campfire flame than further away. But the sun's atmosphere does the opposite. Energy must be bypassing the visible surface of the sun and flowing into the corona.

Much of it appears to come from the transition region – the area between the sun's corona and the chromosphere, the next atmospheric layer down. Tornadoes, rain, magnetic braids, plasma jets and strange phenomena called spicules are all thought to play a role in this heating process, bringing energy from the lower regions of the sun and depositing it higher up. But still no one knows exactly how.

Missions to the sun

To solve some of these solar conundrums, it will help to get as close as possible.

Solar Orbiter is a European Space Agency mission due to launch in October 2018, aiming to fly within 45 million kilometres of the sun. It will photograph the sun's poles for the first time, which should help scientists understand how the sun generates its magnetic field, and may even give insights into why the north and south poles of the field flip so frequently. The probe will also be able to sniff the pristine solar wind before it has reached Earth.

NASA's Parker Solar Probe is set to launch in August 2018 and will get even closer, just 6 million kilometres from the sun's surface. It will approach in a looping, circuitous route, like a matador approaching a wary bull. The slow approach is partly for safety's sake: as the probe gets closer, scientists can carefully monitor any threats from radiation or heat and adjust the approach if anything goes awry. The mission will swing around Venus seven times to put it on the right trajectory. At its closest approach, it will zip past the sun at 200 kilometres per second. It will try to answer questions including how the atmosphere is heated and how the sun generates its wind.

Missing metals

We can't go and take a sample of the sun, but we have two ways to work out what it is made of. Helioseismologists look at vibrations on the sun's surface, which are affected by the chemical composition of the interior. Spectroscopists look at sunlight, passing it through high-tech prisms, decomposing it into stripes that serve as unique barcodes for its constituent elements.

For years, these two methods painted the same picture: the sun is mostly hydrogen and helium, with a sprinkling of other elements donated through the explosions of other stars. Astronomers refer to all these heavier elements as metals. They can be found scattered throughout the interior of the sun, making up a little less than 2 per cent of its total mass. Despite their minority status, these metals play a crucial role, helping to shuttle energy from the core out to the boiling layer on the surface.

Then we hit a snag. In the early 2000s, a young researcher in Copenhagen, Martin Asplund, was studying the motions of the outer layers of stars, a requisite step towards performing more accurate spectroscopic calculations. With a departmental supercomputer at his disposal, Asplund built a three-dimensional numerical model of the sun's outer layer. In 2009 it gave a startling result: a quarter of the metals implied by helioseismology could no longer be found.

So far, nobody has found a way to discount Asplund's conclusions, and as his results have become widely accepted, they've caused ripples well beyond the sun. As our closest and most accessible star, the sun informs our understanding of its cousins across the cosmos.

Some have reached for exotic solutions, suggesting that **dark matter** in the sun could reconcile the results. A much more likely solution is that under the extreme temperatures and pressures of the sun, the heavy elements simply aren't behaving as expected: they may absorb and emit light in a different way.

Our best hope of solving this puzzle is a new Canadian neutrino detector called SNO+. We already routinely detect neutrinos from the sun, but SNO+ may be able to tease out the subtle signal of rare CNO neutrinos – born during fusion reactions involving atoms of carbon, nitrogen and oxygen. Then we could peer directly into the core of the sun to see how many of these heavy elements are in there.

Long-lost solar siblings

The sun may be on its own now – its closest neighbour is 4.2 light years away – but that wasn't always so. Once upon a time it had close family. After their birth in a cloud of dust and gas, these solar siblings scattered hundreds of light years apart. In May 2014, astronomers reported finding the first one: a star called HD 162826 (see Figure 1.2).

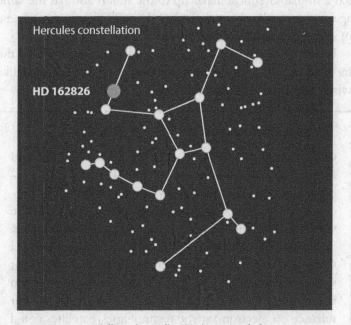

FIGURE 1.2 Modelling the Milky Way's motion led astronomers to HD 162826, which may be a sibling to our own sun.

The star is about 110 light years away, and with the aid of a pair of binoculars you can see it in the left arm of the constellation Hercules. It's just a little warmer and bluer than the sun, and 15 per cent more massive.

To find its family ties, a team led by Ivan Ramirez at the University of Texas at Austin combed through galactic archaeology studies, which model the motions of the Milky Way. These predictions lay out where sibling stars would be now if they had formed in the same place as the sun. Though they spread out in different directions, their positions still give away their birthplace.

Narrowing down the search area to 30 stars, the team looked at them closely to find a family resemblance. Only HD 162826 has a similar chemical make-up to the sun. It also has the same age as the sun. And tantalizingly, HD 162826 is in a catalogue of stars that might harbour planets.

Locating more solar siblings could tell astronomers about the birth of our solar system, including what conditions were like when the sun and planets formed.

The next great space storm

On 2 September 1859, a giant ejection of matter and magnetism erupted from the sun and struck Earth. Auroral storms burst over two-thirds of the planet's skies, compasses went haywire, and the telegraph system across the world malfunctioned as currents surged through the wires.

Named after the British amateur astronomer Richard Carrington – who took observations of it – the event was little more than a spectacular light show to most on the ground. Today it would be a disaster, thanks to our reliance on electromagnetic technologies. Satellites could burn out, and our communications and positioning systems with them. Transformers could be destroyed, bringing down power grids across nations. Public transport would grind to a halt. In 2008, the American National Academy of Sciences estimated that a Carrington event could cost the US economy alone $2 trillion.

Since then, space weather has become a rising priority. The key to protecting ourselves is to understand the sun's capricious magnetism, which is behind coronal mass ejections such as the one that caused the Carrington event. We can't predict when and where these ejections will happen, because exactly how the sun generates its magnetism remains a mystery. As mentioned, the European Space Agency's Solar Orbiter could help change that by measuring solar magnetic fields. If it can unlock the mystery of the sun's magnetic dynamo, it just might help civilization avoid an abrupt return to 1859.

What would happen if a giant comet crashed into the sun?

Most comets that brush past the sun end their lives in a whimper, but if a big enough comet were to plunge straight in, it should go out with a bang.

NASA's Solar and Heliospheric Observatory has detected three or more small comets a week passing very close to the sun. The smaller sungrazers don't usually make it far. It isn't the sun's million-kelvin corona that melts them; that is too thin to transfer much heat. Instead, the intense glare of solar radiation sublimates ices into gas, which escapes into space or causes the comets to crack apart. But some survive. In 2011, comet Lovejoy passed through the solar corona, emerging much worse for wear but still loosely together. Comet ISON barely survived a similar trip in 2014.

So what would happen if a comet hit the sun head on? A team led by John Brown, Astronomer Royal for Scotland, has calculated the answer.

If its course takes it close enough, the steep fall into the sun's gravity would accelerate it to more than 600 kilometres per second. At that speed, drag from the sun's lower atmosphere would flatten the comet into a pancake: a supersonic snowball in hell, as Brown describes it.

Finally it would explode in an airburst, releasing ultraviolet radiation and X-rays that we could see with modern instruments. The crash would unleash as much energy as a magnetic flare or coronal mass ejection, but over a much smaller area. The momentum of the comet could even make the sun ring like a bell, with subsequent sunquakes echoing through the solar atmosphere.

The calculations may also apply to other solar systems, where young stars are bombarded with far more comets than the sun has to face.

2

Worlds of iron and rock

Four small planets occupy the inner reaches of the solar system. The baked, inky surface of Mercury has only recently been mapped, the toxic hell of Venus could hold lessons for Earth, while crowd favourite Mars might yet reveal traces of life. The troubled third planet from the sun is outside the scope of this book, so our lunar companion will stand in for it, being the only other world visited by humans.

Cinderworld

Mercury is an oddball. The chimeric planet has a cratered face like the moon, yet conceals a metal heart larger than that of Mars, making up about 70 per cent of its total mass. It has a remarkably dark surface and a surprising magnetic field. While other planets go around the sun in more or less the same plane, Mercury opts for a jaunty angle; while Earth's orbit is essentially round, Mercury prefers an ellipse.

Many questions remain unanswered because Mercury is the least explored of all the terrestrial planets. NASA's Messenger spacecraft was the first mission to orbit it, from 2011 until 2015, taking 300,000 images and millions of measurements of everything from Mercury's radioactivity to the chemical make-up of its atmosphere. Messenger mapped the planet using a laser altimeter to measure the heights of hills and depths of craters (see Figure 2.1). Its other data are helping researchers tease out some of Mercury's secrets.

A graphite crust

Mercury's surface is exceptionally dark, reflecting much less sunlight than our moon. It was thought that iron and titanium might be to blame, but Messenger didn't find enough of either element. Then a clever analysis of Messenger data acquired right above the dimmest parts of Mercury's surface suggested another answer. A team led by Patrick Peplowski, of the Johns Hopkins University Applied Physics Laboratory in Maryland, combined infrared spectra with the number of neutrons sparked by **cosmic rays** to reveal that the dark stuff is carbon, in the form of graphite.

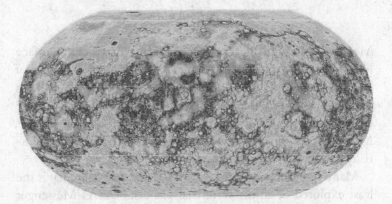

FIGURE 2.1 More than 100,000 images from Messenger combined to create a model of the planet's topography. The highest point is 4.48 kilometres above average elevation, located south of the equator in some of Mercury's oldest terrain; the lowest point 5.38 kilometres below average elevation is at the floor of the volcanic Rachmaninoff basin.

The graphite may date back to the earliest days of Mercury, when a magma ocean covered the planet. Assuming the planet had the same chemistry as it does today, nearly every mineral that formed in the ocean would have sunk to the bottom. The only mineral that would float is graphite. Mercury could have been covered in a shell of graphite a kilometre thick.

Later lava flows would have buried this darker layer. That would mean the darkest material on Mercury today should show up in craters where the original surface has been gouged out – exactly what Peplowski's team found. But this doesn't rule out a rival theory, that comet strikes could have dusted the whole planet with carbon.

Core question

Conventional planet formation models cannot produce Mercury's huge metal core. Astronomers have speculated that Mercury once suffered a massive impact that stripped away most of its rocky mantle – or that its outer layers evaporated away by the heat of the sun. But Messenger found volatile elements such as potassium in the planet's crust. The trauma of either impact or evaporation should have removed those elements.

Meanwhile, observations of extrasolar planets suggest that Mercury's structure is not unique. The two smallest exoplanets whose densities are known, Kepler-10b and Corot-7b, are far denser than expected, suggesting they share Mercury's great heart. These planets, like Mercury, sit close to their sun.

In 2013, Gerhard Wurm of the University of Duisburg-Essen in Germany, and colleagues, suggested a way to explain the whole coterie. Dust grains are heated by starlight, and when gas molecules collide with a hot dust grain, they pick up some of this heat, bouncing off faster than they approached. This gives the grain a little shove. Wurm's group calculated how this photophoretic force would affect dust grains swirling around a star.

Because metallic grains conduct heat well, they are an even temperature throughout. As a result, they will be shoved equally from all sides and so will not move far from the star. But grains that will form rocks, such as silicates, are insulators, so they end up with a hot, sun-facing side, where departing gas molecules will give a bigger shove than those on the cold side. In a forming solar system, this effect could sort the grains, with dense metals left close to the star, and lighter silicates pushed further out. So this process may explain why inner planets like Mercury, Kepler-10b and Corot-7b, are so dense.

The next visitor to Mercury is a Japanese–European satellite called BepiColombo. Due to arrive in late 2025, it could settle this and other questions posed by the iron planet.

What happened to Venus?

It is sometimes called Earth's twin, but in fact the relationship between our home and its nearest planetary neighbour is more Jekyll and Hyde. Venus is the same size and composition as Earth, and gets roughly the same amount of sunlight. It is technically inside the solar system's habitable zone where liquid water can exist – and indeed scientists think that Venus probably once had oceans and maybe life. So how did it become so inhospitable?

Our attempts to find out have been thwarted by opaque sulphuric acid clouds, impenetrable to early orbiters. Of the craft we have sent to investigate the surface, less than half survived the trip, the rest collapsing under the punishing pressures of Venus's atmosphere. The few survivors didn't last long. Between them, they amassed less than a single day of ground observations.

What they saw was a dim, desolate wasteland pitted by endless sulphuric acid rain and scoured by syrupy winds, blowing fast at dawn and dusk but slowing in the heat of the day. If the choking, largely carbon dioxide atmosphere doesn't kill you, the heat – a lead-melting 460°C – surely will.

These scenes support the standard explanation for Venus's predicament: the planet is just a bit too close to the sun. This caused water to evaporate and form a thick atmosphere that trapped heat, leading to a runaway greenhouse effect and today's hellish conditions.

But observations from the Venus Express orbiter cast this simple theory into doubt. In 2007, it spotted ions streaming off the planet. The cause was the solar wind, which sails right through Venus's feeble magnetic field. The wind also triggers regular plasma explosions that rip huge chunks off the planet's atmosphere.

Given this constant assault, not much ancient water would remain in the atmosphere. It may have helped to generate the original runaway greenhouse, but something else must be replenishing the choking atmosphere today – so perhaps that something was important in the distant past too.

The most likely candidate is sulphur and carbon dioxide released by volcanoes at the surface. As yet, no one has found active volcanism on the planet, but evidence is piling up. Venus Express showed that volcanic flows make up 80 per cent of the planet's surface. Some may be just tens of thousands of years old.

Finding out about Venus's past would help us rule out similar dead-end planets in our search for Earth-like worlds around other stars – and maybe tell us whether we have a similar fate in store. Models show that Earth's climate will begin to resemble Venus's in about 2 billion years as the sun ages and slowly heats up. But what if we are actually closer to the edge? Could some unknown variable tip us over the brink of destruction much sooner? Such questions have prompted a flurry of proposals for a return to Venus, hoping to find out whether it was always destined to be an uninhabitable wasteland.

Then again, maybe Venus is not so hostile after all. In the cloud decks 70 kilometres above Venus's infernal surface, the weather is fine: plenty of sunshine and water, and

Earth-like pressures and temperatures. Thanks to these conditions life might exist in the clouds. To find out, we would need some kind of atmospheric rover. Aerospace giant Northrop Grumman has designed an autonomous inflatable spaceplane that could bob around the planet for a year, sniffing for signs of life. NASA's Jet Propulsion Lab (JPL) has a more ambitious concept, an airship that could carry scientists to the balmy clouds of Venus.

Paradise lost

Computer simulations suggest that early Venus might have looked a lot like our home planet, and it might have been habitable until remarkably recently.

David Grinspoon at the Planetary Science Institute in Tucson, Arizona, and his colleagues used a climate model to create four versions for Venus, each varying slightly in details such as the amount of energy the planet received from the sun, or the length of a Venusian day. Where firm information was missing, the team filled in with educated guesses. Venus has an unusually high ratio of deuterium to hydrogen atoms, a sign that it once housed a substantial amount of water, so they added a shallow ocean.

Looking at how each version might have evolved over time, the researchers found that the planet might have looked much like an early Earth, and remained habitable for most of its lifetime, perhaps until about 700 million years ago. The most promising of the four Venuses enjoyed moderate temperatures, thick cloud cover and even the occasional light snowfall. Just like the early Earth, Venus had the requirements for the origins of life – as far as we understand them – says Grinspoon.

He suggests that a future mission to Venus should look out for signs of water-related erosion near the equator, which would provide evidence for the oceans detailed in their simulation. Such signs have already been seen by missions at Mars. NASA is now weighing up two potential Venus projects. One mission would drop a probe through the clouds down to the surface, while another would orbit the planet and image its surface.

The researchers would also like to run simulations of further alternative pasts for Venus – perhaps one where it was a desert world, or submerged in as much water as Earth, to find out which scenario is most likely to lead to the Venus we see today.

Apollo and the birth of the moon

While the world watched in fascination in 1969 as Neil Armstrong and Buzz Aldrin gambolled about on the moon, planetary scientists had their eyes on a different prize. For them, the value of the Apollo 11 mission lay in the cargo it aimed to bring home, and the astronauts did not disappoint. By the time Armstrong and Aldrin climbed into the lunar module for the last time, they had collected 22 kilograms of moon rocks, enough to fill a small suitcase.

Five more Apollo crews brought the total collection of moon rock to 382 kilograms, in some 2200 individually numbered samples. Three uncrewed Russian landers recovered a further 300 grams of soil.

The rocks were billed at the time as a scientific treasure, and they have rewritten our ideas about planetary formation and evolution, laying to rest many myths about the moon. Harold Urey, one of the first great advocates of lunar exploration, had predicted the moon was made of primitive meteoritic material. He was wrong. Instead some of the rocks looked a lot like Earth rocks, notably the dark basalts that give the lunar *maria*, or seas, their distinctive hue. Others were quite different, such as the ubiquitous jumbled pieces of rock called breccias, which have been smashed up and welded together by millions of years of meteorite impacts.

Many clues in the lunar rocks have taken years to decode, and some of the conclusions are still hotly debated. A huge surprise was the evidence that the early moon was covered by a deep ocean of molten rock. The moon's mountainous regions are dominated by anorthosite, a rock rare on Earth that forms when light, aluminium-rich minerals float to the top of a lava pool. If anorthosites are everywhere on the moon, then its entire surface must once have been a magma ocean, and this prompts a puzzling question: what created it?

The prevailing theory today is that it was the result of a cataclysmic event about 50 million years after the solar system began to form, when the Earth was in its infancy. According to this hypothesis, proto-Earth ran into a Mars-sized planet, and debris from the collision entered orbit around Earth where it rapidly coalesced to form the moon.

But recently the Apollo rocks have raised problems for this idea. According to the giant impact hypothesis, some of the moon's material came from the other protoplanet – so you would expect them to be slightly different in composition from Earth rocks, and in particular contain different amounts of isotopes of the same element. But in 2012 Junjun Zhang, then at the University of Chicago, and colleagues, found that the oxygen, chromium, potassium and silicon isotopes in moon rocks are indistinguishable from Earth's.

This has led some scientists to reconsider the theory. Wim van Westrenen, a planetary scientist at the Free University in Amsterdam, has suggested that a giant nuclear explosion inside Earth created the moon. Others have found ways to modify the giant-impact theory, involving a fast-spinning, easily disrupted early Earth and a smaller impactor that could have buried itself deep inside our planet and sent a plume of purely terrestrial rocks into orbit to form the moon.

Another surprise from Apollo samples is on firmer ground. They show that the moon's largest impact craters are roughly the same age, formed 3.8 to 4 billion years ago. The moon, and presumably Earth, must have been subjected to a devastating barrage half a billion years after the solar system formed. To cause this, something big must have been going on back then in the outer solar system, perhaps a shift in the orbits of Neptune and Uranus, sending a stream of comets plunging inwards. Curiously, this episode in the solar system's history, which has come to be known as the late heavy bombardment, ended at about the same time as the first signs of life appeared on Earth. Did it in fact create the conditions under which life could evolve?

The late heavy bombardment, and the giant-impact theory of the moon's formation, have led to a radical re-evaluation of the history of the early solar system. Before Apollo, planetary scientists saw the collection of objects orbiting the sun as a clockwork mechanism in which collisions were rare and insignificant. Now it is recognized as being a far more dynamic environment, in which planets can shuffle around, collide or be ejected altogether. The history of all the inner planets has been shaped by collisions, and nowhere is that history more visible than the moon.

Without the samples brought back from the moon, we might never have made these key discoveries. So do the Apollo rocks harbour any more secrets? The instruments for dating mineral samples have become more sophisticated, enabling researchers to determine the age of ever smaller samples, such as tiny mineral grains within a rock. These techniques have prompted a rethink of some dates in lunar history, dating the formation of the moon to 20 to 30 million years later than we thought, about 4.5 billion years ago, while the last of the magma oceans probably solidified about 4.417 billion years ago.

What the Apollo samples will never do is answer some of the remaining big-picture questions. What will we find on the far side – the half of the moon's surface we can never see from Earth? Can we put together a detailed history of the lava flows that formed the basalts of the lunar seas? Can we find any samples from deep inside the moon? These are all powerful reasons for returning to the moon.

Red rivers

Prospects for life on Mars – past, present and future – hinge on the existence of liquid water. But the waters of Mars are infuriatingly elusive.

Recurring slope lineae are dark streaks that appear during warm seasons, and get longer, and fade in each Martian year. They have long been thought to represent signs of briny water streaming down the sides of craters and hills. Salts can lower the freezing point of water, making it possible for liquid water to exist in theory even in the cold Martian climate.

In 2015 that idea was backed up by data from NASA's Mars Reconnaissance Orbiter, which has a spectrometer that analyses reflected sunlight to detect what minerals are present on the surface. Spectral data from four locations with recurring slope lineae revealed the presence of hydrated salts, which are most likely to be magnesium perchlorate, magnesium chlorate and sodium perchlorate.

Confirmation of flowing water on the surface would add substantial weight to calls for NASA to commit more strongly to searching for life on Mars. In the Atacama desert, one of the most hostile environments on Earth, communities of microbes are able to survive on moisture in the ground created by salts absorbing water from the atmosphere. Some think similar microbes could live on Mars.

But maybe these mysterious dark flows are not water after all? It is hard to melt ice on Mars, or draw it from the thin, dry atmosphere. Instead, the lineae could be rivulets of sand, set in motion by sunlight on the Martian surface.

Frédéric Schmidt at the University of Paris-South, and his colleagues, say the features could be sand avalanches, similar to the ones we might see on a dune on a windy day. Instead of wind, these flows are caused by sunlight and shadow. When sunlight hits the sand, it heats up the top layer while leaving deeper layers cool. This temperature gradient causes a corresponding change in the pressure of tiny pockets of gas surrounding the sand particles, shifting the gas upwards. That in turn jostles grains of sand and soil, causing them to slip down the Martian slopes. This effect should be most pronounced in afternoon shadows cast by boulders or outcrops. Then, the contrast between the cooling top of the sand and the still-warm layers just below creates a second pressure gradient, shifting the gas and sand even more.

Support for the idea that these avalanches do not need water came in late 2017 when further analysis of images taken by the Mars Reconnaissance Orbiter found that the dark lines appear only on slopes steep enough for dry grains to tumble in exactly

FIGURE 2.2 Mysterious lines on the surface of Mars resemble marks
left by flowing water, but could instead be rivulets of sand set in
motion by sunlight.

the way they do on sand dunes. Lead researcher Colin Dundas
of the US Geological Survey Astrogeology Science Center in
Flagstaff, Arizona, says this new evidence supports the idea that
Mars today is very dry.

If recurring slope lineae are created without liquid, it could
dismantle our hopes that they might make life easier, both for
organisms native to Mars and eventual human explorers.

An impossible ocean

As for the past, that is just as baffling. Mars has ice caps, and
liquid water was once copious. Observations of clay minerals
and remnants of lake and river deposits are unequivocal: water
flowed freely between 3.5 and 4 billion years ago, and the
planet probably held a large ocean. There are even signs that
giant tsunamis rearranged the ancient coastline. Alberto Fairén

of the Centre for Astrobiology in Madrid, and his colleagues used thermal imaging from NASA's Mars Odyssey spacecraft to study the boundary between the low-lying Chryse Planitia and the highland Arabia Terra regions of Mars. They identified flows of ice and boulders running uphill and extending hundreds of kilometres across the highlands. These couldn't have been created by gravity-driven processes, but tsunamis could explain them. Simulations suggest an asteroid impact large enough to create a 30-kilometre crater could have generated such a tsunami, which would have been around 50 metres high when it reached the shore.

And yet we can't work out how water could have existed in liquid form on young Mars. This 40-year-old mystery is known as the Mars paradox. If and when we resolve it, we might need to throw away a lot of textbooks.

The trouble starts when you look at the conditions on Mars at that time. Even today, Mars's thin atmosphere and distance from the sun keep it, on average, at about −61°C, cold enough to hold existing water in permanent polar deposits. Billions of years ago, under a younger, cooler sun, it was even colder.

So given that the freezing point of water is the same on Earth as Mars, how was Mars ever warm enough for liquid water to flow? One plausible explanation is that greenhouse gases trapped heat in the way they do on Earth. These gases could have been produced by many sources, including volcanic eruptions. The gas with the best track record of trapping heat is carbon dioxide, but it can't warm Mars enough for liquid water. In any case, sediments laid down 3.5 billion years ago show that the atmosphere then contained only scant amounts of carbon dioxide.

Perhaps the maths could work if you added some methane or hydrogen? No. With so little CO_2, it doesn't matter how much hydrogen or methane or other gases you add into the equation. You need a thick atmosphere to begin with to shield these sensitive greenhouse gases from solar radiation.

How about water salty enough to remain liquid even at water-freezing temperatures? Then the atmosphere wouldn't need much CO_2. But this too could fall short. Ultra-saline water can flow – on Earth at least – but such a cold Mars wouldn't allow enough precipitation to account for the standing water etched into Mars' sandstone and mudstone over millions of years.

So is there some planetary mechanism we still don't understand? A mixture of greenhouse gases we haven't yet hit on? Perhaps the real trouble is our understanding of water itself. We're going to be in truly alien territory when the mystery is solved.

Interview: My year on Mars

Sheyna Gifford spent a year in a dome on a volcano, serving as chief medical officer on a NASA mission that aimed to find out how the first settlers to the Red Planet might get on. New Scientist *interviewed her in 2017.*

How did you and five crewmates end up living on a volcano in Hawaii?

NASA wanted to know what happens, psychologically and socially, when you send six people off to live isolated on another planet – how we work together, how we respond to stress, how well we communicate with Earth. It was their longest-running Mars simulation, called the Hawaii Space Exploration Analog and Simulation (HI-SEAS) mission. For 366 days up to August 2016, six of us lived in a geodesic metal dome.

Who was with you in the dome?

Our group consisted of the commander, science officer, engineer, biologist, architect and me, the medical officer. We wore digital badges that monitored our interactions. For the most part we were lab rats, but we also did our own science, looking at the mountain's geology, testing our hydroponic food-growing techniques and studying our own microbiomes.

What was it like living on 'Mars'?

Our communications with the outside world were all delayed by 20 minutes, just as they could be between Mars and Earth, depending on their relative positions in space. We couldn't use phones or Skype. Every time we went outside the habitat, we wore spacesuits. We didn't see another soul for a year.

Did people fall out with each other?

We had some personality differences, but when times were tough we stayed together because we were all professionals. Working on our assigned duties and focusing on the mission kept the group unified. That was a big lesson.

What challenges stick in your mind?

The hardest part was the second quarter, when we were low on power and food. Due to administrative problems, we didn't get a promised resupply – we were down to two kinds of dried vegetables, spinach and kale, and nobody wanted to eat them. And it was extremely cold. Morale was low, but on a mission like this, there will always be a trough. There was a permanent fallout of our second-quarter blues: we developed a culture of staying in our rooms alone whenever it was cold. That habit never changed.

Did things turn for the better?

We got a resupply at Christmas time that included a third battery, so even when the days were short the lab could run for a day and a half on one day's charging with our array of solar panels. Everything was suddenly great – we turned the heater on and started cooking. On Hannuka I cooked and taught everyone how to play a traditional Jewish game.

How did you celebrate festivals on 'Mars'?

That's an interesting question – what meaning would Christmas have on Mars anyway? It's not connected to Martian seasons or to anyone who ever lived or died on Mars. Instead of Christmas, we celebrated a sort of non-denominational holiday.

But our first Martian holiday was in honour of our first tomato harvest. Our astrobiologist spent months raising those tomatoes. They grew out of bottles, hydroponically, because we had very little soil, just like on Mars. We each got one. We set out plates, sprinkled over dried parsley, lit candles and showed up nicely dressed for our one tomato. We called this holiday the *Jour de la Grand Tomate* – Day of the Great Tomato. That was the first fresh tomato we'd had in at least four months.

I took my tomato and smelled it like a maniac for ten minutes – it smelled like a whole hothouse of tomatoes. When I finally tasted it, it burned my lips. There wasn't anything wrong with it, there was something wrong with my lips. We didn't have any acidic food; we had been eating powdered tomatoes. I had to eat the tomato carefully.

Did you miss family and friends?

Hugely. Getting email from them was essential to my well-being. I watched as crewmates who did not have that support suffered. Just having that email arrive is critical — it's proof that you still exist and still matter.

Based on your experiences, what sort of outlook would the first Mars colonists develop?

If you took someone born and raised on Mars and dropped them in Times Square, they would freak out at the amount of electricity being used for no good reason. Probably all the electricity we produced in a day would be burned in seconds. Earthly trash cans are full of things we Martians would never throw away. We either reuse it or melt it down and 3D-print it into something else. We don't value stuff on Mars except in terms of its utility. Money is useless, and the only thing that matters is how smart, sane and capable you are.

What was it like being the chief medical officer?

The job of a doctor gets redefined. You go back to being like the old town doctor who goes around talking to people about their health, trying to keep them from getting sick. Once they get sick, what you can do is limited.

A space doctor begins each day with a prayer that they won't have to do their job that day. We were lucky — only one person got seriously hurt during the mission, and it was me. I was exploring the local terrain when a lava tube collapsed on me and I hurt my knee.

You also experimented with virtual reality. What was that all about?

NASA want to know if VR can help with some of the loneliness or boredom when you're living in a distant dome. They created an immersive experience on Earth using a 360-degree camera and recorded audio, and then we could step into it using equipment in the dome. The VR took me to Boston. I put the goggles on, and I was suddenly standing back on a familiar street. People were looking at me – they had been recorded approaching the VR camera, gesturing at it. It was like I had been physically beamed back to Earth. That would be a great way to use VR on Mars.

The day I participated in that experiment, my grandmother passed away – expectedly. The communication delay couldn't be turned off because it wasn't a crisis situation under the rules of our mission. So I had to say goodbye to her over delayed video, which is something you never want to do.

Some might argue it's better to send people with less to lose into space. Would you agree?

The question is more basic: do you send social or asocial people? I say send people with the largest number of earthly attachments, for several reasons. One, if crew relationships go to hell, they'll turn back to their support network. Two, they will fight like crazy to get back; they will hold the ship together. Three, which might be the most important reason: people on Earth want to come too, but they can't. So you want super-social citizens of the world who constantly send messages back. We should send people who have lived in multiple countries, have practised multiple religions and are as flexible as possible. You're not up there for you, after all. You're up there for Earth.

A planet among asteroids

Ceres is both the largest asteroid in the solar system and the smallest known **dwarf planet**, a terrestrial outpost on the rim of the inner solar system. NASA's Dawn probe arrived in orbit in April 2015, and immediately discovered perplexing white spots on this little world. Probing their nature may provide clues to Ceres's interior.

In 2016, Dawn took a close look at the biggest of these bright areas, which sits within the 92-kilometre-wide Occator crater. The probe saw a central mound, and colour variations across the surface of the bright area that wouldn't show up to the human eye, but reflect possible differences in the composition of the material. How this arrangement formed is still a mystery. One idea is that a meteorite hit Ceres to expose icy material from as much as 40 kilometres below the surface, while also heating it up. As this material settled into the Occator crater we see today, the water would have evaporated, leaving bright salt and minerals behind.

The floor of Occator is also riddled with fractures that seem to be older than the crater itself. These could have provided a handy route for material beneath the surface to be squeezed out when the crater was formed.

Meanwhile Dawn found water ice hiding in a 10-kilometre wide crater called Oxo. The water is near the side of the crater and could be in one of two possible forms: either ordinary ice, or locked up inside hydrated minerals. Models of Ceres's formation, along with the sightings of bright spots like the ones at Occator, suggest the dwarf planet has an icy sub-surface layer that is mixed up with salt and rock. Oxo's ice could have been exposed by a landslide, or dug up by a meteorite hitting the surface. Normally on Ceres this ice would boil away, leaving behind bright salts, as is thought to be the case at Occator. But Oxo provides a cool place for ice to hang on.

In 2017, planetary scientists found pockets of carbon-based organic compounds on the surface, forming tar-like minerals. Their composition can't be pinned down precisely, but their spectral fingerprints match the make-up of the tar-like minerals kerite or asphaltite. The constituents and concentrations of these organic materials suggest that it's unlikely they came to Ceres from another planetary body.

They wouldn't have survived the heat of an impact, and if they had hitched a ride on another stellar object, they would be widely dispersed, rather than concentrated in pockets. So they must have come from Ceres itself.

Along with the discoveries of water ice and bright mineral deposits, this points to a more complex picture of the dwarf planet than we once assumed. What it's doing on the inside is not entirely clear yet, but the organic material on the surface indicates that there are processes within Ceres regulated by heat and water.

Could violent sandstorms threaten astronauts, as in *The Martian*?

In *The Martian* (2015) a monstrous sandstorm threatens to blow over the rocket that the crew intends to use to get home. In real life this is unlikely, because the atmosphere of Mars is so thin — only 1 per cent of the density of Earth's. Even unusually high winds on Mars would produce a force comparable to a fresh breeze on Earth.

But dust storms on Mars can still be dangerous: they impair visibility and reduce your ability to harvest solar energy. This has been a serious problem for Mars rovers Spirit and Opportunity.

Dust devils could cause mischief too. The dust grains would be hard to keep out of spacesuits, shelters and just about everything else, and the static electricity from grains rubbing against one another in these dry, sandy whirl-winds could cause trouble.

In 2017 a team led by Brian Jackson at Boise State University in Idaho analysed barometer data from the Martian surface to show that dust devils are much more common than we thought. The team found that on any given day about one 13-metre-wide dust devil pops up per square kilometre of surface, on average. If you were standing on the surface, you might be able to see dozens of kilometre-tall dust devils skittering across the ground at any one time.

A history of the heavens

Astronomy is by far the oldest science. For many thousands of years it has been an aid to navigation and timekeeping, and astronomical alignments are built into many ancient monuments.

~3500 BCE +
Written records of astronomy begin with the Sumerians, whose base-60 number system gives us the units of angle we still use to mark out the sky.

~3000 BCE +
Chinese astronomers develop their own methods, build detailed star catalogues and record astronomical phenomena including eclipses, sunspots and novae.

1543
Nicolaus Copernicus presents his sun-centred model of the solar system.

1054
Chinese astronomers record the supernova that created the Crab nebula.

~1570–1601
Tycho Brahe makes the most precise observations yet of the planets and other objects. He shows that comets and novas are not atmospheric phenomena, but distant objects, challenging the idea that the celestial realm is unchanging.

1609
Johannes Kepler publishes two of his three laws of planetary motion, which are based on Brahe's and his own observations. The first law, that planetary orbits are elliptical, overturns an ancient assumption that they are based on perfect circles.

1781
William Herschel discovers Uranus, the first new planet since antiquity.

1705
Edmond Halley uses Newton's theories to work out that certain bright comets seen in previous centuries were actually appearances of the same comet on a long, highly elliptical orbit, returning to the inner solar system within a period of about 76 years.

~250 BCE
Eratosthenes measures the
circumference of the Earth.

~140 CE
In Alexandria, Ptolemy refines the
Earth-centred (geocentric) model
of the solar system, adding epicycles
to the motions of the planets.

~800
The golden age of Islamic
astronomy begins.

499
Indian mathematician Aryabhata publishes
a great treatise on astronomy, which among
other things explains the causes of eclipses
and gives a highly accurate length of the year.

1610
Galileo points a telescope at Jupiter and observes its four large moons.
This increases the known number of moons in the universe fivefold,
and shows that heavenly bodies do not have to orbit the Earth,
supporting the heliocentric model of Copernicus. Among many other
astronomical observations, Galileo also discovered Saturn's rings.

1687
Isaac Newton lays out a physical theory of the heavens.
His equations describing the force of gravity and its
effects can be combined to explain and precisely
predict the motions of planets, moons and other objects
in space – a field known as celestial mechanics.

A history of the heavens (*continued*)

1814
With his invention, the spectroscope, Josef von Fraunhofer sees dark lines in the spectrum of light from the sun. Decades later scientists realise that these betray the presence of different atoms. Spectroscopy enables astronomers to work out which chemicals make up stars, planets and interstellar clouds.

1938
Hans Bethe shows that nuclear fusion is the main source of power in most stars, as Eddington had suggested.

1924
Edwin Hubble finds that – as others had suspected – many nebulae are actually other galaxies beyond the Milky Way. This enormously expanded our view of the universe. A few years later he showed that the universe is actually expanding.

1967
Jocelyn Bell and Antony Hewish discover pulsars – spinning ultra-dense neutron stars created by supernovae.

1970s and 80s
The two Voyager spacecraft give us out first close-up views of Jupiter, Saturn, Uranus and Neptune, and some of their remarkable moons.

1821
Alexis Bouvard suggests that the gravitational disturbance of an unknown planet is affecting the position of Uranus.

1846
Johan Galle is the first person to see the planet proposed by Bouvard. He finds it using Urbain le Verrier's calculations of its likely position. The planet is named Neptune.

1924
Arthur Eddington calculates a relationship between mass and luminosity based on his pioneering physical model of stars, which also implies that stellar cores reach temperatures of millions of degrees.

~1910
Ejnar Hertzsprung and Henry Norris . Russell plot stars by colour and luminosity, revealing different classes of star and hinting at their evolution.

1994
Comet Shoemaker-Levy 9 hits Jupiter.

1995
Michel Mayor and Didier Queloz find evidence for a planet circling the main-sequence star 51 Pegasi. A flood of exoplanet discoveries follows.

2016
The LIGO collaboration publish the first direct detection of gravitational waves, from two colliding black holes.

2005
The Huygens probe lands on Titan and gives us our first view of the giant, smog-bound moon's surface.

3
Among the giants

The four giant planets of our solar system are remarkable enough, with their vast storms, supersonic winds, strange cloud patterns and unfathomed depths – but the many moons that orbit them are perhaps even more fascinating, harbouring Stygian oceans, volcanoes, geysers and methane rainstorms.

King of the planets

Jupiter outweighs all the other planets of the solar system put together. Its gravity shapes the asteroid belt. Its powerful magnetic field contains savage radiation belts. And magnanimously it hoovers up rogue comets that might otherwise make life on Earth more perilous.

In July 2016, NASA's Juno spacecraft went into orbit around Jupiter, in a mission aimed at unravelling how the planet formed. Juno's nine instruments have studied Jupiter's structure, peering deep into the gargantuan storms that churn across the planet. The probe has mapped the planet's gravity and magnetic fields, looking for evidence of a solid core, and observed auroras undulating across the atmosphere.

Juno is the first spacecraft destined to orbit over the poles of a gas giant, rather than around the equator. This unusual trajectory protects it from the worst of Jupiter's radiation belts, which could permanently fry its electrical systems.

After a mission lasting just 37 orbits, Juno met its end with a dramatic flourish early in 2018, when it deliberately crashed into Jupiter.

In April 2017, early data from Juno were already challenging assumptions about everything from the planet's atmosphere to its interior. They revealed a dense zone of ammonia gas around Jupiter's equator, plus other regions where ammonia is depleted, which together suggest an ammonia-based weather system. We have long known that Jupiter is shrouded in ammonia clouds, but the existence of such a deep belt of the gas, plunging 300 kilometres below the cloud, was surprising – suggesting a weather system that penetrates much deeper than anyone expected.

Another shock was that Jupiter's magnetic field is stronger and much more irregular than expected. The irregularity may be a sign that the dynamo driving it is very diffuse and occurs much further from the core than previously thought, perhaps in a layer of metallic hydrogen.

The early orbits also produced several new insights into the planet's atmosphere. The probe's camera sent back amazing pictures of dozens of huge cyclones over the poles, each hundreds of miles across and hitherto unsuspected. Strange white ovals

FIGURE 3.1 Swirling patterns above Jupiter's north pole as captured by Juno.

were spotted in belts south of Jupiter's equator. They could be clouds containing ammonia and hydrazine, a substance used as rocket fuel on Earth.

Dissolving core

The probe also challenged models of what's inside the planet. We had assumed Jupiter has a fairly uniform interior. The molecular hydrogen atmosphere goes down around 1000 metres. Below that, the pressure is thought to crush hydrogen into a metallic state: protons meandering through a sea of electrons. Here, drops of helium and other elements may be raining down from the atmosphere above. Finally there seems to be a smaller solid core around 70,000 kilometres down. This picture was based on mapping the planet's gravity.

Initial gravity measurements from Juno suggest that the internal layers are not completely regular in their make-up, and point to a core that is not solid like Earth's, but fuzzy – edged and mingled with the overlying metallic hydrogen layer. That could fit with calculations from 2011, which show that Jupiter's rocky core may be dissolving like an antacid tablet plopped in water.

Giant planets like Jupiter and Saturn are thought to begin their lives as solid bodies of rock and ice. When they grow to about ten times the mass of Earth, their gravity pulls in gas from their birth nebula, giving them thick atmospheres made mainly of hydrogen. Curiously, some studies have suggested that Jupiter's core may weigh less than ten Earths, while the core of its smaller sibling Saturn packs a bigger punch at 15 to 30 Earths.

Some researchers had proposed that the intense pressures and temperatures at Jupiter's heart might cause its core to dissolve into the surrounding atmosphere, which is at such high pressure that it behaves somewhat like a liquid. Hugh Wilson, now at the CSIRO in Melbourne, and Burkhard Militzer of the University of California, Berkeley, used the equations of quantum mechanics to see how the mineral magnesium oxide – thought to be a constituent of Jupiter's core – behaves at Jupiter-like pressures of about 40 million Earth atmospheres and temperatures of 20,000°C. They found that magnesium oxide does indeed dissolve into its fluid surroundings in those conditions. The dissolved rock might get mixed into the rest of the atmosphere over time.

In Saturn, which is about a third of the mass of Jupiter, conditions are not as extreme, and so if Saturn's core is dissolving, the calculations suggest it will be happening slowly. Meanwhile the process probably happens much more rapidly in planets more massive than Jupiter, and many large exoplanets (see Chapter 7) may have no cores at all.

Land of lava lakes

Pockmarked with sulphurous pits, bathed in intense radiation and shaken by constant volcanic eruptions, Io is the fiery hell of the solar system. Despite being cold enough in parts to be covered in layers of sulphur dioxide frost, this large inner moon of Jupiter is the most volcanic world known, spitting out 100 times as much lava as all Earth's volcanoes can muster, from a surface area just one-twelfth the size.

Io's surface is dotted with bubbling lakes of molten rock, the largest of which, Loki Patera, is more than 200 kilometres across. Elsewhere, magma suddenly forces its way out of fissures in the rocky crust, creating lines of lava fountains that can stretch for 50 kilometres or more. NASA's New Horizons spacecraft picked up the heat from one of these great curtains of fire in 2007 as the probe passed by Jupiter *en route* to Pluto.

Some of Io's eruptions are violent enough to hurl giant plumes of gas and dust 500 kilometres into space. This can happen when a lava flow vaporizes the surface layers of frozen sulphur dioxide, or when bubbles of gas form inside rising magma to blast high-speed debris out through the moon's surface.

All this volcanic violence results from a tug of love between Jupiter and Io's two siblings, Europa and Ganymede. These moons have orbital periods exactly two and four times as long as Io's, which results in the three moons lining up every so often. Over time, the gentle gravitational tugs of this periodic conjunction have gradually nudged Io into an elongated orbit. As Io moves around this orbit, the grip of Jupiter's gravity weakens and strengthens, flexing the moon's rock. These stresses and strains warm the moon from within in a process called tidal heating. This effect is so powerful on Io that it can melt rock, creating the volcanoes.

In 2013, researchers looked back over snapshots of three of Io's hotspots: Pillan, Wayland Patera and Loki Patera, taken by the Cassini probe when it flew by in late 2000. By working out the temperature of the lakes, Daniel Allen at Lake Land College in Mattoon, Illinois, and colleagues, determined that the lava in all three lakes was most likely to be molten basalt.

They also found that each one has its own eruption style. Pillan is the architect of the three. Previous probes saw it erupt in 1997, spewing out enough lava to cover 5600 square kilometres. Cassini's temperature readings suggest it is now surrounded by a relatively tall mass of cooling rock that has built up around the lava lake. Wayland, meanwhile, is a bit of a burnout. Roughly 95 kilometres across, it appears to be either a cooling lava flow or a lava lake during a period of low activity, says Allen. Then there is Loki, the trickster. It is huge, spanning 200 kilometres, and emits around 13 per cent of all the heat from Io. Depending on when you visit, you might find a solid crust, potentially capable of supporting a heat-proofed rover, or a molten morass, or even glorious lava fountains.

Such extreme volcanism may be common in the universe. The exoplanet COROT-7b, for example, orbits very close to its star and so feels a very strong gravitational pull. If its orbit is only slightly elliptical, there will be enough tidal heating to plaster the planet with volcanoes. So Io may be giving us a glimpse of conditions on a million hellish exoplanets.

Io itself seems to be cooling, probably because its orbit has become less elliptical than it once was. Tens or hundreds of millions of years from now, the orbital resonance with Europa and Ganymede is likely to grow out of sync, letting Io settle into a nearly circular orbit with almost no tidal heating. Then Io's fires will finally fade.

Deep dark seas

'Follow the water' has long been the mantra in the search for life, because every known organism needs water to survive. Most prospecting has been done on Mars, but its water is either long gone or locked in the ground as ice.

In contrast, Jupiter's moon Europa and Saturn's Enceladus both have deep liquid water oceans beneath their frozen outer shells. Astrobiologists wonder whether they might hold alien echoes of the extreme ecosystems on our own ocean floor, where life is fuelled by nothing more than the reaction between rock and water. The race is on to spot signs of similar geochemical rumblings on Europa and Enceladus, and discover whether we are alone in the solar system.

The first hints of Europa's concealed sea came from the Voyager probes, which explored Jupiter in the 1970s. Voyager II spotted cracks in Europa's icy surface crust, suggesting active processes below. When the Galileo spacecraft returned in the 1990s, it saw another clue: Jupiter's magnetic field lines were bent around Europa, indicating the presence of a secondary field. The best explanation is a global vat of electrically conductive fluid, and seawater fits the bill. We now think this ice–enclosed ocean reaches down 100 kilometres. If so, it contains enough salty water to fill Earth's ocean basins roughly twice over.

The case for a sea on Enceladus washed in more recently. In 2005, the Cassini probe showed that the moon leaves a distinct impression in Saturn's magnetic field, indicating the presence of something that can interact with it. That turned out to be an astrobiologist's fantasy: a plume of ice particles and water vapour shooting into space through cracks near Enceladus's south pole.

FIGURE 3.2 Plumes of ice particles and water vapour shoot into space through cracks near the south pole of Enceladus.

Cassini has since flown through these plumes several times. First its instruments revealed the presence of organic compounds. Particles collected from the lowest part of the plumes were rich in salt, indicative of an ocean beneath. Cassini detected ammonia, too, which acts as an antifreeze to keep water flowing even at low temperatures. All the signs suggested this was a sea of liquid water, stocked with at least some of the building blocks of life.

The treasures kept coming. In March 2015, Cassini scientists detected silicate grains in the plumes – particles that most likely formed in reactions at hydrothermal vents. By September, measurements of how Enceladus's outer crust slips and slides had convinced them that it contains a global ocean between 26 and 31 kilometres in depth. That's a paddling pool compared with Europa's, but way deeper than Earth's oceans.

So when can we visit? NASA plans to send a mission to Europa in June 2022. It will feature a magnetometer to probe the ocean's saltiness and ice-penetrating radar to show where solid shell meets liquid water. It might even include a lander to fish for amino acids, the building blocks of the proteins used by every living thing on Earth.

NASA has also invited proposals for a trip to Enceladus. One option is the Enceladus Life Finder, a probe that will sample plumes using instruments capable of detecting larger molecules and more accurately distinguishing between chemical signatures. Other plans have even suggested carrying samples back to Earth for analysis.

With any luck, probes will be arriving at these ocean worlds by the twilight years of the 2020s. Meanwhile there is plenty we can do to plumb Europa and Enceladus's hidden depths. We can survey their surfaces using ground-based telescopes, looking at the fissures where water might bubble through and leave telltale deposits from the oceans beneath. We can model the geophysics that keeps them liquid so far from the sun, and may generate conditions that could support life. And we can use the closest analogues on our own planet to guide our search.

On Earth, deep-sea vents at the boundaries between tectonic plates, where magma breaches the sea floor, have long been recognized as hotbeds for life. In the lightless depths around geysers of scalding, murky water known as black smokers, bacteria feed on chemicals, and all manner of organisms make their living on those microbes. Europa or Enceladus might draw enough energy from the tidal push and pull of their host planets to have molten interiors that can fuel similar vents.

The good news is that we now know of another possibility. At the Lost City vents beneath the Atlantic, discovered in 2000, a hydrothermal ecosystem thrives without the faintest rumble of tectonic activity. Lost City is powered by a chemical reaction called serpentinization. When alkaline rocks from Earth's mantle meet a more acidic ocean, they generate heat and spew out hydrogen, which in turn reacts with the carbon compounds dissolved in seawater, acting as food for microbes. According to Michael Russell, a geologist turned astrobiologist at NASA's Jet Propulsion Lab in Pasadena (JPL), California, Lost City is just the sort of place where life on Earth might have begun.

To find out if this is happening on Enceladus, the Cassini team have been looking for hydrogen in the plumes. The spacecraft did detect hydrogen in early passes, but there was no way to determine if it came from the moon itself or from inside the instrument, because when particles from the plumes entered the spacecraft's mass spectrometer they interacted with its titanium walls, producing hydrogen. So the team had to put their instrument in a new mode that measured the molecules without allowing them to touch the walls. Finally, with data gathered on Cassini's last pass through the plumes, they found the molecular hydrogen they were looking for – and a lot of it. There is too much hydrogen to be stored in tiny Enceladus's ice shell or ocean, so it must be continuously produced there, probably by hydrothermal reactions.

Europa is also likely to have serpentinization, and it is much larger than Enceladus, meaning it boasts more rock in contact with seawater. In 2016 Kevin Hand at JPL, and his colleagues published a study suggesting that Europa's ocean has a chemical balance similar to Earth's. The calculations were based on estimates that fractures in the moon's sea floor could reach as deep as 25 kilometres into the rocky interior. In that case, there would be great swathes of rock surface with which water can react to release lots of hydrogen.

For life as we know it, electron-grabbing oxidants like oxygen and electron-giving reducing agents like hydrogen have to meet and react, releasing energy that living things rely on in the form of electrons. Europa has no atmosphere from which to cycle oxygen, as Earth does, but we know that radiation from Jupiter produces oxidizing chemicals on its surface. Hand and his colleagues assumed that these oxidants are being cycled from surface to sea, an assumption that might be tested by a crust-sounding seismometer onboard a future Europa lander.

It is quite possible, of course, that life elsewhere follows a different rulebook, and is made from a different set of building blocks. So what should we be looking for if not organic molecules and amino acids? It is a question that astrobiologists contemplate, but it can probably only be answered by finding alien life forms.

If we can detect something akin to deep-sea alkaline vents on faraway moons, the odds of finding extraterrestrials would be slashed. We might also have to entertain the prospect that similarly life-friendly conditions are lurking beneath the shells of other icy worlds, such as Jupiter's giant moons Callisto and Ganymede, or the dwarf planet Ceres. We now know that oceans concealed by frozen crusts are common in the solar system. Perhaps they are the default condition for life, while our blue planet, with its peculiar open oceans, is the outlier.

Ring master

Saturn is the jewel of the solar system. Its beautiful rings make the second largest planet unique. It is the most diaphanous of the planets, being less dense than water; and the least spherical, spinning so fast that it is visibly flattened at the poles.

The Cassini mission arrived in 2004 to turn its panoply of instruments on this marvellous world, overturning much of what we thought we knew, and discovering surprising new features of the ringed planet such as the giant hexagonal jetstream that kinks its way around Saturn's north pole. It also toured the moons, exploring the two-faced Iapetus, strangely corroded Hyperion, geyser-bearing Enceladus and the giant moon Titan with its methane lakes and rivers.

Echoes from the deep

Even Cassini could not see directly beneath Saturn's colourful bands of cloud. But in 2015, it did pick up some tantalizing clues about the planet's interior. Disturbances in the planet's system of rings pointed to planet-wide tsunamis racing around the equator, and hinted at surprising structures inside – perhaps giant whirlpools thousands of kilometres deep, a buried sphere of light or something even stranger.

In 1980, the first Voyager mission found that Saturn's rings are home to spiral-shaped density waves, looking a little like the arms of a spiral galaxy. Most of these waves radiate outwards, and they are known to be caused by the gravity of Saturn's moons. But a few waves move inwards, and researchers suspected that they are echoes of much more substantial waves deep inside the planet.

According to conventional wisdom, Saturn is a uniform fluid ball, a smooth mixture of hydrogen and helium; and in theory, this substance can form waves that race around the equator. The undulating gravity of the peaks and troughs on one of these planetary waves could be enough to tease out a spiral wave in the rings above.

But based on the limited data from Voyager, no one could be sure. So Phillip Nicholson and his colleagues at Cornell University started picking through observations gathered by Cassini, which orbited Saturn until 2017. They traced out several spirals in the inner rings which back up the basic idea: waves are indeed racing around within the fluid body of the planet. Then things got weird.

If the planet is a simple fluid ball, the theory goes, the speed of each wave should be fixed by its number of peaks. A three-peaked wave travels more slowly than a two-peaked wave, and so on. The researchers expected to see one example of each type of spiral, each whizzing around at a unique speed. Instead, Nicholson's team found three separate three-armed waves, all travelling at slightly different speeds, as well as two separate two-armers.

One explanation would be a large solid core, vibrating in its own way, interfering with the simple fluid waves above. While that's in line with conventional ideas of planet formation, it would take some fine-tuning to generate these particular waves.

Alternatively, there may be a layer in the planet where the hydrogen–helium mixture behaves oddly. At some point the molecules of hydrogen and helium should break up into separate atoms, which would make the mixture relatively transparent – creating a luminous sphere. This would vibrate differently. If so, the spirals could be telling us something about what happens to matter under these pressures, a regime still beyond our computer simulations.

Strangest of all, a few spiral waves are moving around at almost exactly the same speed as Saturn's rotation. One explanation is the presence of permanent hills and valleys on the planet. But if Saturn is fluid, this would be like finding fixed hills on the sea.

Unless the laws of physics have been repealed on Saturn, fluid hills aren't an option. Nicholson's colleague Maryame El Moutamid has another tentative suggestion: massive vortices deep inside the planet, less dense than the surrounding fluid and so exerting less gravity. This would create dents in Saturn's gravitational field to explain those perplexing spirals.

The circle's beginning

Saturn's spectacular rings are made up of trillions of icy pieces in thousands of fine strands, shepherded by tiny moons. Propeller-shaped patterns in the rings are helping some researchers to study the origin of planets in the dusty discs around young stars. But the origin of the rings themselves remains elusive.

One popular idea is that rings form when a passing asteroid or comet is pulverized by the gravity of a planet. But that doesn't explain why Saturn's rings are mostly water ice, while rings around other gas giants are rocky.

FIGURE 3.3 Vibrations in Saturn's rings suggest strange goings-on beneath the planet's colourful clouds.

In 2016, Ryuki Hyodo at Kobe University in Japan and his colleagues built a new model of ring formation. They considered the way the passing object whirls through space: whether its tumbling lines up with the direction in which it travels around the planet, or if it is doing backflips. The team found that bodies spinning in the same direction as their path around the planet are more easily broken up, and their fragments more efficiently sucked into orbit. That is because the planet's gravity pulls harder on the closer side of the small object, tugging it around in the same direction as it is travelling. If the planet's gravity has to work against the object's spin, it will be unable to sweep in as much material as when they are aligned.

The team then simulated what Saturn and Uranus might do to passing objects spinning in different ways. They modelled more complex bodies than have been tried before: rather than just a homogeneous ball, they included more realistic objects with a hard, rocky core surrounded by an icy mantle.

In some Saturn scenarios, only the outer layer of frozen water was swept up by the planet, creating proto-rings that could have evolved into the icy bands visible today. The Uranus simulations, however, tended to produce rockier rings. Because Uranus is denser than Saturn, its gravity can seize more of the deeper, rockier part of a passing body.

A puzzling question remains. Saturn and the other giant planets would have been most likely to encounter passing bodies in the tumult of the early solar system about 4 billion years ago. Since then, most of those objects have smashed into planets or been ejected from the solar system. But the clean water ice of Saturn's ring system suggests that it is much younger than that, since interplanetary dust should pollute it over time. And observations by Cassini point to a low mass for the rings, making them less robust and relatively unlikely to last for billions of years. Was Saturn simply lucky to meet a passing ice world in the recent past, or did it shred one of its own moons after some orbital shenanigans? Or are the rings ancient after all, simply left over from the planet's formation, somehow surviving unpolluted?

The seas of Titan

The sky is a baleful orange, but then it's always like that. The roar of the approaching maelstrom is new though, and disturbing. As are the gathering clouds, which threaten to loose a deluge more violent than anything ever seen on Earth. A sailor on an alien sea may hesitate: is it wise to venture into the Throat of Kraken?

One day this could be a real scene on Saturn's giant moon, Titan. It is the only moon of the solar system with a thick atmosphere, and apart from Earth the only world known to have liquid on its surface. While rivers of water run through mountains of rock on Earth, on Titan the streams are liquid methane and the hills and plains are made of iron-hard water ice.

Long before the Cassini mission arrived, scientists had calculated that methane and other liquid hydrocarbons might collect into seas – perhaps even forming a global ocean. Nobody could be sure, though, because an orange layer of smog hides the moon's surface. So Cassini carried a lander, the Huygens probe, that was designed to float.

When Huygens did plunge through the smog in 2005 it sent back images of an eerily Earth-like landscape. The probe landed on a pebbly mudflat. The ground there was soaked with methane, but it was hardly the hoped-for ocean.

Our first clear sight of lakes came in 2006, revealed by Cassini's radar. This instrument was able to peer through the smog, to map a thin slice of the moon's surface each time Cassini passed by. As the picture built up over the years, we saw that some of these bodies of liquid were large enough to be considered seas. The largest, named Kraken Mare after the monster of Norse legend, is about 1000 kilometres long.

On 23 May 2013, the spacecraft flew low over Ligeia Mare, the second sea of Titan. Its radar was aimed straight down, enabling the instrument to trace out the height of land and sea by sending out sharp pulses of radio waves, then measuring how long it took for the reflections to ping back.

When the team first looked at the data there was a ping from the sea surface, as expected – and then a second faint ping, not much more than a microsecond after the first, bouncing off the seabed. This was the first time we have plumbed the depths of any sea or lake beyond Earth. The

FIGURE 3.4 Radar has plumbed the depths of Ligeia Mare, Titan's second sea, and revealed what it is made of.

timing of the second reflection shows that Ligeia Mare is about 160 metres deep.

It came as a surprise that the seabed can be seen at all. Titan's atmosphere is choked with complex hydrocarbon molecules that absorb radar, and everyone assumed that some of these would muddy the seas. So see-through Ligeia must be free from those complex hydrocarbons. Some blend of ethane and methane seemed to be the most likely option.

Because methane should evaporate quite rapidly from the seas, planetary scientists expected lingering ethane to dominate, but new lab data showed that ethane absorbs the radio waves too strongly to fit the seabed sighting from Titan. So something must be refreshing the northern seas with crystal-clear methane – perhaps seasonal rainstorms like those that soaked huge tracts of land near the equator in 2011.

Cassini has also seen a strange feature in Ligeia Mare that disappears and reappears. Some researchers have suggested that this 'magic island' could be a raft of nitrogen bubbles fizzing out of the methane.

Release the Kraken

As the largest alien sea, Kraken Mare holds a particular fascination, and most plans for a maritime mission to Titan have targeted it. Kraken Mare is cut almost in two by promontories of land and a string of islands. Ralph Lorenz at the University of Arizona in Tucson named this feature the Throat of Kraken – and he realized that something special might be going on at this maritime divide. The gravity of Saturn is expected to cause tides in the seas of Titan. As the moon follows its orbit, the tide should rise and fall in Kraken Mare by around a metre. As it rises in the north it should ebb in the south, flowing from one end of the sea to the other, funnelled through the Throat.

There, the tidal current could reach speeds of 2 kilometres per hour, according to Lorenz's calculations. While that may seem slow, on Titan gravity is much lower than on Earth and the liquid in the seas is much lighter, so even such a gentle tidal current may be brisk enough to send the Kraken into a frenzy. Lorenz likens the Throat to the west coast of Scotland, where two islands frame the Strait of Corryvreckan. Tidal currents here churn the sea surface into a maelstrom, with one of the largest whirlpools in the world. Could the Kraken host a whirlpool too? If so it would be a curious coincidence, as legend tells that the monster after which the sea is named would create whirlpools to drag sailors to their doom.

Icy life

As well as mythical monsters, could the seas of Titan host real, living creatures? If so, they would have to be quite unearthly. On Earth, each living cell is a packet of mostly water surrounded by a thin membrane made of lipids, and neither of these components would fare very well on Titan, which is far too cold for liquid water, with average surface temperatures of −149°C.

But in 2017 Maureen Palmer at NASA's Goddard Space Flight Center in Greenbelt, Maryland, and her colleagues reported finding traces of vinyl cyanide in the moon's nitrogen atmosphere. According to a 2015 study, vinyl cyanide is particularly good at forming the stable, flexible structures necessary to build something like a cell membrane. There is a lot of vinyl cyanide on Titan, according to Palmer's results. That much building material means a higher likelihood that membranes could grow large enough to support complicated structures like cell innards.

Membranes alone wouldn't be enough, of course, but high in Titan's atmosphere Cassini has detected a molecule called a carbon chain anion that might help life along. Ravi Desai at University College London and his colleagues think that these anions may form the seeds for larger, more complex organic molecules closer to the surface.

A Titanic voyage

Bays and beaches a billion kilometres from Earth; a view of Saturn's rings rising above the waves and whirlpools; an exotic chemistry that could illuminate the origins of life. Titan could hardly be a more alluring destination, and some planetary scientists have drawn up plans for a space boat, or even a submarine, to explore its distant seas.

At the Southwest Research Institute in San Antonio, Texas, Hunter Waite developed a submarine mission concept in 2010. His floating mothership holds a submersible that would dive down by letting methane flood a hollow chamber. Later the sub can discard the chamber and ascend to the surface again.

There might be organic sediments at the bottom of the seas, holding chemical treasure – especially if liquid water from Titan's interior is seeping out down there, mimicking the oxygen-free, organic-rich environment of the early Earth. A submarine could also measure the isotopic mix of various chemicals, to help geologists learn how Titan formed and evolved.

Sideways

Most of the planets in our solar system rotate around roughly similar axes, spinning in the same plane as their orbit. Uranus is the odd one out. It spins on its side, tilted almost 98 degrees from the plane of its orbit around the sun.

It also has the strangest magnetic field in the solar system. The axis of the field is tilted at a 59-degree angle from the rotational axis; it is off-centre, with the field lines emerging about a third of the way towards the south pole; and whereas Earth's magnetic field resembles that of a bar magnet, on Uranus nearby patches of the surface can have fields of opposite polarity.

As happens with many other planets, the magnetic field of Uranus blows a bubble around it called a magnetosphere. A model published in 2017 suggests that the edge of this magnetosphere could be slamming open and shut every day.

The magnetosphere acts as a barrier to the solar wind. When the two are moving in the same direction, the solar wind slides off it like water off a duck's back. But just as when water hits a duck's feathers from the tail end, the duck gets wet, so when the solar wind blows toward Uranus at the right angle, the planet's magnetic field lines up with the solar wind's and lets some particles flow through.

This process, called magnetic reconnection, occurs occasionally near Earth's poles, where the influx of particles from the solar wind can lead to intensified auroras. Carol Paty at the Georgia Institute of Technology in Atlanta and her student Xin Cao modelled the same process on Uranus, and found that it should happen every single day (roughly 17 Earth hours), switching the magnetosphere's protection on and off. This could lead to an aurora there as well.

Weird water

Some of the strange features of the magnetic field could be explained by an equally strange form of water. Simulations in 1999 and an experiment in 2005 hinted that at very high pressures and temperatures, water might behave like both a solid and a liquid. The oxygen and hydrogen atoms in the water molecules would become ionized, with the oxygen ions forming a lattice-like crystal structure and the hydrogen ions able to flow through the lattice like a liquid. This 'superionic' water, forming at temperatures above 2000°C or so, should glow yellow.

The conditions that exist deep within both Uranus and Neptune could be ideal for superionic water to form. In 2010 new computer models, created by a team led by Ronald Redmer of the University of Rostock in Germany, suggested that both planets possess a thick layer of the stuff. The simulations assume the most extreme conditions possible inside both planets, with temperatures reaching up to 6000°C and pressures 7 million times the atmospheric pressure on Earth. The results show that a layer of superionic water should extend from the rocky core of each planet out to about halfway to the surface.

That tallies nicely with the results of a 2006 study led by Sabine Stanley, now at the University of Toronto, and Jeremy Bloxham of Harvard University, attempting to explain why the magnetic field is so patchy on Uranus – and also on Neptune.

Stanley and Bloxham's work suggested that the interiors of both planets contain a narrow layer of electrically conducting material that is constantly churning, which generates magnetic fields. This conducting layer would be made of ordinary ionic water, in which the molecules have broken down into oxygen and hydrogen ions. The study also indicated that the convecting zone cannot extend deeper than about halfway down to the planets' centres. If it were thicker, it would produce a more orderly field like that of a bar magnet.

The transition from convection to non-convection at the depth calculated by Stanley and Bloxham might seem irrelevant, since the superionic water takes over here. But superionic water also conducts electricity, via the flow of hydrogen ions. So something must be stopping the superionic water from churning and making the magnetic field more orderly.

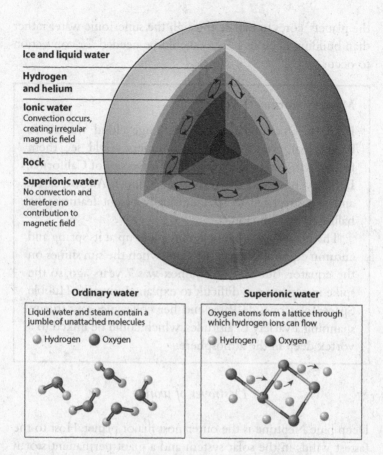

Ice and liquid water

Hydrogen
and helium

Ionic water
Convection occurs,
creating irregular
magnetic field

Rock

Superionic water
No convection and
therefore no
contribution to
magnetic field

Ordinary water

Liquid water and steam contain a
jumble of unattached molecules

● Hydrogen ● Oxygen

Superionic water

Oxygen atoms form a lattice through
which hydrogen ions can flow

● Hydrogen ● Oxygen

FIGURE 3.5 A layer of superionic water seems to lie around the rocky cores of both Neptune and Uranus. It may not undergo convection, which could help explain the planets' strange magnetic fields.

One possibility is that superionic water is mostly transparent to infrared radiation, or heat. The electrons in superionic water can absorb infrared radiation, but simulations indicate they tend to stay near the oxygen atoms, making most of the space transparent to heat. That would make it easy for heat from the planets' cores to radiate through the superionic water rather than building up at its base, as would be needed for convection to occur.

Mystery storms

In 2015, Uranus played host to huge cloud systems so bright that even amateur astronomers could see them from Earth. Imke de Pater at the University of California, Berkeley observed the planet on 5 and 6 August, 2014, and was surprised to spot unusually bright features, the hallmark of high clouds.

The planet's weather generally picks up at its spring and autumn equinoxes every 42 years, when the sun shines on the equator. But the last equinox was 7 years ago, so the spike in activity was difficult to explain. Using the Hubble Space Telescope, de Pater and her colleagues saw storms spanning a variety of altitudes, which could be linked to a vortex deep in the atmosphere.

Destroyer of worlds

Deep blue Neptune is the outermost major planet. Host to the fastest winds in the solar system and a giant permanent storm known as the great dark spot, it seems to have had a turbulent past too. In 2010, researchers suggested that Neptune may have swallowed a super-Earth (a large rocky planet) and stolen its moon to boot. The brutal deed could explain mysterious heat radiating from the icy planet and the odd orbit of its moon Triton.

Neptune's own existence was a puzzle until recently. The dusty cloud that gave birth to the planets probably thinned out further from the sun. With building material so scarce, it is hard to understand how Uranus and Neptune, the two outermost planets, managed to get so big.

But what if they formed closer in? In 2005, a team of scientists proposed that the giant planets shifted positions in an early upheaval. In this scenario, Uranus and Neptune formed much closer to the sun and migrated outwards, possibly swapping places in the process. This upheaval could have launched the hail of comets, known as the late heavy bombardment, that scoured the inner solar system.

It would also mean there was enough material just beyond the birthplace of Neptune and Uranus to form a planet with twice the Earth's mass, according to calculations by Steven Desch of Arizona State University in Tempe. They suggest that Neptune's peculiar moon Triton may once have been paired with this hypothetical super-Earth.

Triton is a giant moon that moves through its orbit in the opposite direction to Neptune's rotation, suggesting that it did not form there but was captured instead. For that to happen, the moon would have had to slow down drastically. One way to do this is for Triton to have had a partner that carried away most of the pair's kinetic energy after an encounter with Neptune.

In 2006 researchers argued that Triton was initially paired with another object of similar size that wound up being gravitationally slung into space after the pair ventured near Neptune. But Desch calculated that Triton could have slowed even more if its former partner was a heavy super-Earth, able to carry away more of the pair's kinetic energy.

Neptune may have engulfed the super-Earth. Heat left over from such a huge impact could explain why the planet radiates much more heat than its cousin Uranus.

Are Neptune and Uranus made of ice?

These planets are often called ice giants, but they don't contain ice in the everyday sense of the word. Planetary scientists use the word to refer to compounds that freeze at the typical temperatures of small objects in the outer solar system, such as comets. While Jupiter and Saturn are mainly made from hydrogen and helium, the interiors of Neptune and Uranus are thought to be mainly water, ammonia, methane and other 'ices'. They aren't frozen solid. At the high temperatures and pressures in there, these compounds are in a fluid form.

4
The wild frontier

Beyond Neptune, the ice worlds swarm. This was once a little regarded zone of the solar system, but we now know it as a vivid and diverse realm – thanks to newly discovered dwarf planets and the remarkable images from the New Horizons mission to Pluto. There are also hints of something much bigger out there, a large planet apparently lurking beyond our sight.

Pluto unveiled

For 85 years since its discovery in 1930, Pluto remained little more than a faint dot to observers on Earth. So nobody knew what to expect when NASA's New Horizons spacecraft beamed back the first close-ups in July 2015.

The images stunned researchers. The dwarf planet is unlike any other world we've ever visited. Under a surprisingly complex atmosphere are floating mountains, suspected ice volcanoes and cracked terrain like the highlands of Mars. Most shocking were smooth regions on the surface. Before the probe's arrival, researchers expected to see a heavily cratered world, bombarded since their formation in the early days of the solar system. But the smooth terrain shows that it has been geologically active.

Strangest of all is Sputnik Planum, a 1000-kilometre plain divided up into rough polygons a few tens of kilometres across. These are convection patterns; they show that the nitrogen ice of Sputnik Planum is churning, like the surface of the sun or oil in a pan.

While this slo-mo maelstrom is certainly bizarre and unexpected, it is not as hard to explain as was first thought. Nitrogen ice is not only soft, but also an excellent thermal insulator, meaning even a feeble heat source from below can build up the temperature, kicking off convection. The small stock of heat left over from Pluto's turbulent formation, supplemented by heat from the decay of radioactive trace elements within its core, is expected to add up to about 4 milliwatts per square metre – enough to drive the convection of Sputnik Planum.

Strange pockmarks are peppered across this living, shifting landscape, possibly caused by sublimation of the nitrogen ice. And blocks gathered at the junctions of some convection cells (see Figure 4.1) are probably hills of water ice floating on the denser nitrogen. On the north-western flank of Sputnik

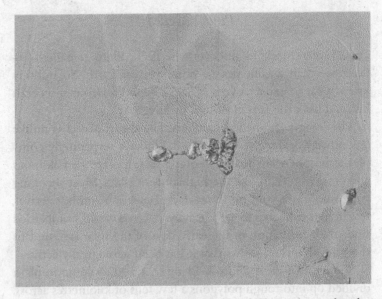

FIGURE 4.1 Floating ice hills on Pluto's Sputnik Planum, here gathered into a formation informally identified by researchers as a Klingon 'bird of prey' spacecraft.

Planum, ice blocks gather in the jumbled peaks of the al-Idrisi range. These kilometre-high mountains may also be floating, or may have become beached.

Nitrogen snow and ice volcanoes

Pluto's nitrogen-dominated atmosphere is cold and thin, with a pressure at ground level equal to that 80 km above Earth. Hazy layers of fine aerosol particles stretch up to 200 km above the ground.

The weather seems to be surprisingly like Earth's, but with a nitrogen rather than a water cycle. Nitrogen sublimates from the ices of Sputnik Planum, like water evaporating from Earth's oceans. It then falls as snow or freezes out as frost on the eastern highlands, finally flowing back down to the plain in glaciers. There are even signs of nitrogen fog in places – and perhaps even clouds.

Wright Mons is unlike most of the jagged mountains that punctuate Pluto's surface. A hummocky mass with a huge central pit, it looks suspiciously like a volcano. If so, then 4-kilometre Wright Mons and its even taller neighbour Piccard Mons would not have erupted molten rock, but instead some chillier fluid – probably water mixed with another substance that lowers its melting point.

Wright Mons is no relic from the dwarf planet's early days. Its sides bear hardly any visible impact craters, so can't have been exposed for too long to the rain of space debris recorded elsewhere on Pluto's surface. It is probably much younger than a billion years and perhaps only a few million. So is it extinct, or merely dormant?

The case of the missing craters

Across both Pluto and its large moon Charon there is a relative shortage of large craters. That may be telling us something profound about how planets form.

According to the traditional picture, small bodies called planetesimals grew in the early days of the solar system as little rocks gradually came together to make bigger rocks. This process should produce a lot of objects a few kilometres in diameter, and far fewer objects tens or hundreds of kilometres across.

Planetesimals of all available sizes should hit Pluto and Charon from time to time, forming craters. So the relative lack of smallish craters on Pluto seems to paint a decidedly non-traditional picture. It might support an alternative model called pebble **accretion**, in which large planetesimals form almost instantly when swarms of little pebbles immersed in gas suddenly collapse. This may be a vital stage in building not just little icy worlds like Pluto, but also the cores of gas giants and warm rocky planets such as Earth.

Canyonland

Charon is half the diameter of Pluto, making it the biggest moon relative to its host in the solar system. Some researchers consider them twin dwarf planets.

Before New Horizons, we suspected that Charon would be a dull, monotonous world, but it has a colourful

FIGURE 4.2 Pluto's giant moon Charon bears a vast a belt of fractures and canyons.

and varied landscape. Just north of the equator is a belt of fractures and canyons stretching for nearly 2000 kilometres. The system is four times as long as the Grand Canyon and twice as deep in places. It hints at a violent period in Charon's history when the crust was torn open.

To the south is a smooth plain, dubbed Vulcan Planum. It has fewer large craters than the north, suggesting that the surface was laid down relatively recently. One possibility is that an internal ocean froze and cracked the crust, allowing lava to reach the surface.

Ice swarm

Along with Pluto, thousands of icy bodies are known to orbit beyond Neptune. Most of these **trans-Neptunian objects** (TNOs) occupy the **Kuiper belt**, a flattened disc stretching between about 30 and 50 **astronomical units**, or AU (one AU is Earth's mean distance from the sun).

Other TNOs have more tilted and elongated orbits, forming what's known as the scattered disc. This is thought to be the origin of short-period comets, as these orbits can be destabilized by gravitational nudges from the giant planets, occasionally sending an object into the inner solar system where the heat of the sun turns ancient ice into a brilliant **coma** and tail.

One object, discovered in 2016, orbits in a plane that's tilted 110 degrees to the plane of the planets, swinging around the sun backwards. The discovery team nicknamed it 'Niku' after the Chinese adjective for rebellious – and it's unclear what could have knocked it so far out of the usual orderly plane of the solar system.

Most known TNOs are tens to a few hundred kilometres across, and New Horizons is due to visit one of these, 2014 MU69, in January 2019. This is a member of the 'cold classicals' group, which move in relatively circular orbits compared with other Kuiper belt objects, and tend to be reddish in hue. New Horizons's encounter with 2014 MU69 will be fleeting, but it should help to reveal whether Pluto was formed from such objects and could help to address fundamental questions about how planets form. Objects like 2014 MU69, as residents of a region that has remained largely undisturbed since the early days of the solar system, are thought to be the ancient leftovers from the planet-forming process.

Observations could also help to answer how the solar system came to be arranged in its particular way. Existing models suggest that the gas giants were once bunched up much more tightly than they are today, encircled by a substantial disc of planetesimals. Then something destabilized this cosy arrangement, hurling the planets into their present-day positions. The outer disc was shaken up too, although some of it endured to form the Kuiper belt.

Was this a violent upheaval or a more gentle migration? To answer that, we need to know how massive the disc of planetesimals was before the gas giants moved. Here is where 2014 MU69 could help. If it is riddled with impact craters, for example, that would suggest there were once lots of objects around to crash into it. So by studying the craters in New Horizons's images, scientists should get a better idea of the mass of the disc.

A few denizens of the Kuiper belt and scattered disc are large enough to be considered dwarf planets. Pluto, Eris, Haumea and Makemake are already recognized as such by the International Astronomical Union; many more probably will be before long. They all have moons – in fact, after discoveries in 2016

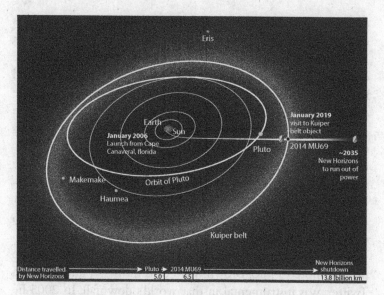

FIGURE 4.3 Before it falls silent forever, the New Horizons probe is heading into the Kuiper belt for one last mission: a close encounter with a pristine remnant from the early days of the solar system called 2014 MU69.

and 2017, all ten known TNOs with diameters near or above 1,000 kilometres are known to have at least one moon – suggesting that they share a crowded, chaotic past. They have a wide range of colours and shapes, with Haumea for example being stretched into an extremely elongated ellipsoid. Haumea also has a ring. Might some of these dimly glimpsed worlds be as complex and active as Pluto turned out to be?

Planet nine… and ten?

Is a planet ten times the mass of Earth lurking in the outer reaches of the solar system? In 2014, astronomers realized that the orbit of a newly discovered TNO called 2012 VP113 was strangely aligned with a group of other objects. Two years later, Konstantin Batygin and Mike Brown of the California Institute of Technology in Pasadena studied these orbits in detail, and found that six follow elliptical orbits that point in the same direction and are similarly tilted away from the plane of the solar system. They suggested that something must be shaping this alignment – and according to their simulations, that something could be a planet that orbits on the opposite side of the sun to the six smaller bodies. The presumed planet's high, elongated orbit takes it from around 200 AU right out to about 700. It would take between 10,000 and 20,000 Earth years just to complete a single orbit.

If the planet is confirmed, it could steal the crown lost by the now-demoted Pluto. Brown, who calls himself 'plutokiller' on Twitter, was instrumental in that world's downfall. In 2005 he discovered a Pluto-sized object, now known as Eris, leading to the reclassification of both objects as dwarf planets.

The elongated trajectory has led some to suggest that it was once an exoplanet and was kidnapped by the sun. If so, it might explain why the other major planets are out of line with the sun. They circle the sun in a plane that is canted by 6 degrees relative to the sun's own equator, an offset that might have been caused by the gravity of the highly-inclined Planet Nine.

However, computer simulations by Richard Parker at the University of Sheffield in the UK and his colleagues suggest otherwise. In their simulated star-forming region, very few free-floating planets get captured. Instead, it is possible that Planet Nine was pushed out from the central solar system when the gas giants reshuffled their orbits.

If Planet Nine really is out there, with any luck astronomers will be able to spot it soon. Slight perturbations in Saturn's

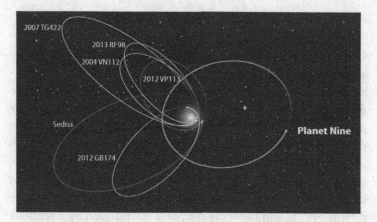

FIGURE 4.4 The alignment of the orbits of six trans-Neptunian objects suggest the presence of a ninth planet.

orbit, if caused by the new planet, would suggest it could be found towards the constellation Cetus (next door to Aries and Pisces). By coincidence, this small zone is already being scoured by the Dark Energy Survey, a project designed to probe the acceleration of the universe.

There may be a second undiscovered planet beyond Neptune. In 2017 Kathryn Volk and Renu Malhotra at the University of Arizona uncovered signs of a weird warp in the Kuiper belt. Across most of the belt the average inclination of orbits matches the plane of the known planets, as expected, but Volk and Malhotra found that at the outer edge of the belt beyond 50 AU, orbits are canted by 8 degrees on average relative to that plane. An unobserved planet with a similar mass to Mars at around 60 AU could cause this warp... but others are doubtful that a planet so close and so large would have remained unseen.

Out to the Oort

Comets are agglomerations of dust and ice that orbit on highly elliptical paths, acquiring their spectacular tails in the head-wind of charged particles streaming from the sun. Some come from the scattered disc of TNOs, tugged from their regular orbits by Neptune and Uranus. These comets can have orbits of no more than 200 years. Other comets, such as Hale-Bopp which flashed past Earth in 1997, have much longer orbits, taking them far further from the sun than anything known in the scattered disc. The conclusion is that they have a more distant origin, and the solar system is surrounded by a tenuous halo of icy outcasts, thrown from the sun's immediate vicinity billions of years ago by the gravity of the giant planets.

This celestial Siberia is known as the **Oort cloud**, after the Dutch astronomer Jan Oort who proposed its existence in 1950. It has never been seen, but if the longest-period comets are anything to go by, it must be vast, reaching out perhaps 100,000 AU (15 trillion kilometres, or 1.6 light years). At such huge distances, it would not be passing planets that throw the comets sunwards – it would be the tug of the Milky Way and nearby stars. The Oort cloud would be where our solar system meets the void.

In 2003, Mike Brown of the California Institute of Technology in Pasadena and his colleagues, observed a dwarf planet, Sedna, swinging on a highly elliptical orbit reaching out to about 1000 times the Earth–sun distance. That gave succour to the idea that within the sphere of the Oort cloud's lies a disc of objects in the plane of our solar system, sometimes known as the Hills cloud.

Oort cloud objects are thought to be stuff left over from when the planets formed, and getting an idea of how many of them there are at different sizes could help us understand how that process occurred. But that's not easy. So far, our information about this primordial rubble comes from stray comets and observations of the largest Kuiper belt objects, which should have a similar composition. The numbers and trajectories of the long-period comets seen so far suggest that the Oort cloud contains trillions of objects a kilometre across or larger, with a combined mass several times that of Earth. That is more material than our current ideas about the solar system's formation can explain – which means that our models might need a fundamental overhaul.

Interview: To Pluto and beyond…

Alan Stern is an engineer and planetary scientist at the South-west Research Institute in Boulder, Colorado. He is principal investigator for NASA's New Horizons mission to Pluto (and beyond). New Scientist *interviewed him in 2016.*

What's the best thing the New Horizons probe revealed about Pluto during its fly-by?

That it is so amazing. This small planet seems to have a little something for everyone – from mountains to blue skies, active geology to glacial fields, and many other types of terrain, geology and activity. It also has a big satellite system. It's a scientific wonderland.

Looking deep into space, do you think there may be a large planet in the Oort cloud – the region far beyond the Kuiper belt?

Absolutely, 100 per cent. I don't believe we can escape that idea.

How quickly could a probe get to a planet in the Oort cloud?

First we would have to find it, which is a very challenging problem. And remember, the Oort cloud is about a hundred times further from the sun than Pluto is. New Horizons, the fastest spacecraft ever launched, took a decade to get to Pluto. The Oort cloud, by present-day technology, is a thousand-year journey.

When you launched New Horizons, Pluto was a planet, but soon after was demoted to 'dwarf planet' status. How do you feel about that?

The International Astronomical Union decided to create a definition that would limit the number of planets specifically so that schoolchildren wouldn't have to memorize too many names. I don't find this particularly scientific. In the outer solar system, we see objects that have the attributes of planets, but we shouldn't worry about how many there are – just like the stars and galaxies. I debated with an astronomer on US national radio who followed the IAU line. He said, 'My little daughter can't possibly remember the names of 50 planets.' I replied, 'Then I guess we're going back to eight US states.'

What specifically is wrong with the IAU's definition of a planet?

The IAU imposed a criterion that a planet has to control its orbital zone – that is, clear it of other objects. But these zones get bigger the further you go from the sun: the zone in which Pluto orbits is larger than those of all the other planets put together. If you put Earth where Pluto orbits, it wouldn't, by that definition, qualify as a planet.

One of your initiatives, Golden Spike, involves selling tickets for commercial trips to the moon. How's that going?

It's a large enterprise, putting together human expeditions to the moon. It's going slower than we initially expected. But this isn't unique to Golden Spike: all commercial space-flight companies are running late. Take suborbital flights. Development began in 2004 and paying customers were expecting to fly into space within a few years. The first tourists are still waiting, 12 years later.

Why are you such a passionate advocate of private and commercial space exploration as opposed to government initiatives?

Government space agencies lead at the very frontier: they develop the technologies and techniques, and are generally first to orbit or make landings on other planets, for example. But they have limited resources. To really become a spacefaring civilization, we need more than just one way to reach space. Private industry creates a big multiplier effect.

How are you involved with Blue Origin, the space-travel company created by Amazon founder Jeff Bezos?

During the four years or so that New Horizons was between Jupiter and Pluto, I consulted for a large number of commercial space companies and universities. Jeff Bezos hired me to help Blue Origin with early efforts to use its New Shepard launch vehicle for research and education purposes. I hope to travel in the New Shepard's crew capsule.

What about your work with Virgin Galactic?

Virgin hired me to help with the development of research and education. Subsequently, in my day job at the Southwest Research Institute in Boulder, Colorado, we've developed a programme of human suborbital flights, with Virgin Galactic as one of our two flight providers – the other being XCOR's Lynx spacecraft. We've arranged three flights on Virgin to run separate biomedical, remote-sensing and microgravity experiments. Three researchers will be flying, myself included.

You also founded Uwingu, a company that for a fee allows people to name features on Mars and newly discovered exoplanets. What happens to the money you raise?

We turn the proceeds into grants for space organizations and researchers, and even space graduate students. We're very proud of being able to take the public's interest and create a 'triple win': the public is more engaged; it creates a revenue stream for our company; and this creates grants for space organizations and individuals.

Have people shown much interest in naming bits of Mars?

There are half a million unnamed features on Mars. In the two years that we've been involved, almost 20,000 features have been named. We've shown that not only do people really love doing this, but by involving the crowd, we make progress much faster in coming up with a complete map of Mars.

The IAU – traditionally the arbiter of naming solar-system objects and features – has recently launched its own public scheme for naming exoplanets. How do you feel about that?

Many IAU members have told me that Uwingu catalysed their activity. There are about 160 billion planets in the galaxy and only 7 or 8 billion people on Earth, so there will be plenty of planets to go around.

Why can't we see the Oort cloud?

The typical object out there is probably just a few kilometres across, and existing in almost total darkness. It is simply too dim and distant for our telescopes to detect. But Oort cloud objects should block and diffract the light coming from distant stars, which astronomers could use to measure their size and distance. Flickers induced by turbulence in Earth's atmosphere make the detections of Oort objects impossible from ground-based detectors, but future space-telescope surveys should be able to detect them in great numbers.

5
The life of stars

Our galaxy harbours hundreds of billions of stars. No two are quite the same. Some are bright, others faint; some are blue, others white, yellow, orange or red; some are enormous, others tiny; some are newborn while others are old and dying. Solving the puzzle of starlight was one of the great triumphs of the last century, but there are plenty of strange stars out there that we still don't understand.

The stellar spectrum

To make sense of the tremendous diversity of stars, astronomers use the Hertzsprung–Russell (H–R) diagram – published by Danish astronomer Ejnar Hertzsprung in 1911 and independently by US astronomer Henry Norris Russell in 1913 (see Figure 5.1). Just as the periodic table allows chemists to sort the elements by their fundamental characters, so the H–R diagram allows us to distinguish stars by their main features. It plots two basic stellar properties: **luminosity** and colour.

Luminosity is just the technical term for brightness: the amount of light and other radiation that a star emits. If the most luminous star in the galaxy replaced the sun, Earth's oceans would boil and its rocks melt. Conversely, if the least luminous star replaced the sun, daytime would be darker than a moonlit night and our oceans would freeze. On the H–R diagram, the most luminous stars appear at the top and the least luminous at the bottom. Because its luminosity is in the middle of this range, the sun appears about halfway down.

To the untrained eye, all stars may look white or yellow. But in fact, stars range in colour from blue and white to yellow, orange and red. This colour tells us how hot the visible surface of a star is. Orange and red stars are between 2000°C and 5000°C, yellow stars such as the sun are 5000°C to 7500°C, and blue and white stars between 7500°C and 50,000°C. On the H–R diagram, the hot blue stars appear on the left hand side, the warm yellow stars in the middle and the cool red stars on the right. Because the sun is yellow, it again lies near the middle of the diagram.

The temperature of a star determines its broad colour and the lines in its spectrum from different types of atom and molecule – which astronomers use to classify its spectral type. For example, white stars have strong spectral lines due to hydrogen,

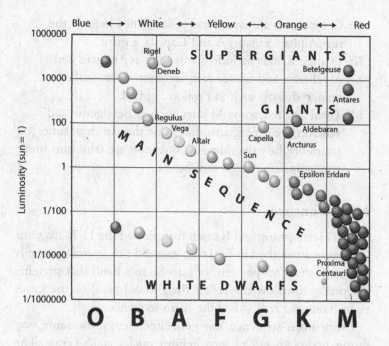

FIGURE 5.1 The Hertzsprung–Russell diagram distinguishes stars by luminosity and colour.

whereas yellow stars have strong lines due to calcium. The main spectral types are:

O: These stars are the hottest and bluest of all.

B: Many bright stars in the sky, such as Spica, Regulus and Rigel, are B-type stars.

A: A-type stars are white and contribute much to the light of our galaxy. They include Sirius, Vega and Altair, which are all main-sequence stars, and Deneb, a white supergiant.

F: The F stars are yellow-white. Viewed from Earth, the two brightest F-type stars are Canopus and Procyon. Polaris, the Pole Star, is also an F star.

G: G-type stars are warm and yellow. They include the sun, Alpha Centauri A and Capella, a giant.

K: Orange K stars include giants such as Arcturus and Aldebaran, and fainter main-sequence stars (called orange dwarfs) such as Epsilon Eridani.

M: Cool and red, some M stars, such as Betelgeuse and Antares, are supergiants that shine thousands of times more brightly than the sun, but most are faint stars on the main sequence – red dwarfs.

Main sequence

When Hertzsprung and Russell first plotted the H-R diagram, they were astonished to find that stars did not scatter randomly over it. Instead, 95 per cent of stars lie in a band that stretches diagonally from the upper left (bright and blue) to the lower right (faint and red), called the main sequence.

Every main-sequence star generates energy the same way, fusing hydrogen nuclei into helium nuclei at its centre. The greater the mass of a main-sequence star, the hotter the star's centre and the faster the hydrogen fuses, so the hotter, bluer and brighter the star.

Yellow main-sequence stars have about the same mass as the sun. Blue and white ones have more – the largest is more than 100 times the sun's mass. Orange and red main-sequence stars have less, down to only 0.07 solar masses.

Less than one star in a thousand is a blue main-sequence star. These massive stars are rare both because few such stars are formed and because they do not live for long, burning their hydrogen fuel at a frantic rate. The most massive stars use up the hydrogen fuel at their centres just a few million years after they are born. Many such stars are visible to the naked eye, because they are so luminous that they can be seen from great distances.

In fact nearly all the stars visible to the naked eye are more luminous than the sun.

In contrast, less massive stars abound but are hard to see. The most abundant main-sequence stars are the red dwarfs, which appear at the bottom right of the H-R diagram. Red dwarfs burn their fuel so slowly that some will remain on the main sequence for thousands of billions of years which is one reason they are so numerous. They outnumber all other stars put together, accounting for 75 per cent of the stars in the galaxy. Yet they are so faint that not a single one is visible to the naked eye.

If a star has even less mass than a red dwarf, it never becomes hot enough to sustain hydrogen fusion and so never joins the main sequence. These stars are called brown dwarfs.

When a main-sequence star uses up the hydrogen at its core, it begins to burn hydrogen in a shell around the core, and then helium within the core. The core shrinks while the rest of the star expands and cools. The star has now left the main sequence to become a giant or supergiant.

Big and bright

Most giants and supergiants are warm to cool, appearing upper right on the H-R diagram. A few are blue or white, such as Rigel, a blue supergiant, and Deneb, a white supergiant.

In general, supergiants have evolved from the hottest and bluest main-sequence stars, whereas giants have evolved from less massive main-sequence stars.

Because of their size, giants and supergiants emit a lot of light. When our sun becomes a giant, it will be about 100 times brighter than it is now. Giants and supergiants are rare because they don't last long. Supergiants soon run through the available fuels – first fusing helium, and then carbon, neon, oxygen, silicon and sulphur, the last two of which fuse into iron. Each stage yields less energy, and iron would actually require energy to fuse into heavier elements. With the internal heat source exhausted, the core collapses to form a **neutron star** or **black hole** (see Chapter 6). This implosion generates so much energy it blasts away the outer layers of the star in a supernova explosion.

Few stars go through this ordeal, because most are born with less than eight solar masses. After becoming red giants, these less massive stars blow away their outer atmosphere into space, exposing a hot core too small to collapse into a neutron star. Radiation from this core makes the ejected atmosphere glow as a **planetary nebula** – so named not because it has anything to do with planets but because through a small telescope it may look like a planet.

Magnitude, distance and luminosity

In about 120 BCE, Hipparchus classified the stars into six groups, from first magnitude (the brightest as seen from Earth) down to sixth magnitude (the faintest). The system was defined more exactly in the 1850s with a logarithmic scale so that each magnitude corresponds to a factor of 2.5 in brightness. A star with an apparent magnitude of 1 is 2.5 times brighter than a star with an apparent magnitude of 2. Most of the brightest stars in the night sky are of the first magnitude; the faintest stars that the naked eye can see are sixth magnitude.

A star's apparent magnitude depends on how far away it is, and distances are often given in light years. One light year is the distance that light travels in a year, or 9.5 million million kilometres. This is an enormous distance: there are as many Earth–sun distances in a light year as there are inches in a mile (more than 60,000). Yet even

---

the nearest star to the sun is 4.24 light years away, and most stars that you can see in the night sky are a few hundred light years away.

Once they know a star's distance, astronomers can calculate its luminosity from its apparent magnitude. This can be expressed as a power output in joules per second, or compared with the sun – a number of solar luminosities – or as absolute magnitude. This is the apparent magnitude the star would have if it were 32.6 light years (10 parsecs) from Earth.

Fading stars

In only a few tens of thousands of years, the planetary nebula cast off by a red giant dissipates, leaving behind a small but extremely hot star: a **white dwarf**. A typical white dwarf is little larger than the Earth but contains about 60 per cent of the mass of the sun. A teaspoonful of white dwarf matter would weigh more than a tonne.

Because so many stars become white dwarfs, these objects are common, making up 5 per cent of all stars in the galaxy. But they are so faint that all are invisible to the naked eye.

A typical white dwarf leads a boring life. It no longer burns fuel; it shines simply because it contains a store of heat. As it radiates energy into space, the star fades and cools over billions of years. Despite their name, white dwarfs can in fact be any colour. The newest are hot and blue, while those that have been around a long time and lost most of their energy are orange or red. So on the H–R diagram, white dwarfs form a sequence that is parallel to the main sequence. If enough time elapses, a white dwarf will fade completely and become a black dwarf. But no black dwarfs exist yet, because the universe is not old enough.

On rare occasions, white dwarfs can create spectacles. If another star orbits the white dwarf and dumps material onto it, the material can explode. Astronomers then see a nova, during which the star increases in brightness some 100,000 times. Violent though it may be, the explosion does not destroy either star.

However, if a companion star transfers too much mass, so the white dwarf reaches 1.44 times the mass of the sun, carbon and oxygen suddenly fuse in a nuclear detonation, annihilating the white dwarf in what's known as a type 1a supernova.

The extremes of temperature and pressure in a supernova can forge lots of iron. Because the explosion destroys the star, all of this iron escapes into space. Together with debris from planetary nebulae, supernova material eventually gathers in star-breeding areas, where it will give birth to new stars and planets, some of which may one day support life. This is how the sun and Earth formed 4.6 billion years ago. We are part of this legacy: apart from the hydrogen, almost all the atoms in our bodies were created by stars.

Metal light

While hydrogen and helium form the bulk of stellar material, most stars also have a healthy dash of heavier elements (which astronomers confusingly call 'metals') inherited from earlier stellar generations. Not so SDSS J102915+172927, which is around 4000 light years away. It is an almost pristine blend of hydrogen and helium, with just 0.00007 per cent other stuff.

That is similar to the primordial matter emerging from the big bang. Such pure gas, lacking the carbon and oxygen that normally help clouds to cool and condense, was thought to form only colossal, short-lived stars. No one knows how this anomalous object managed to form – perhaps it was a fragment spun off during the birth of a supergiant star, back in the dark ages of the universe.

Starbirth

Stars form within dense, dark clouds of molecular hydrogen gas that collapse under their own gravity. This may be triggered when clouds collide or when a stellar explosion nearby sends a shockwave through the cloud. As the cloud becomes fragmented, the first bright new stars often light up remaining filaments of gas, as in the famous 'pillars of creation' in the Eagle

FIGURE 5.2 The Hubble Space Telescope revealed three giant tendrils of dust in the Eagle nebula, cradles of star formation.

nebula, which the Hubble Space Telescope observed in 1995 and then again in 2015, finding that a few tendrils of gas had shifted in the interim. When the gas finally clears from such star forming regions, they leave behind an open cluster such as the Pleiades, before each star drifts off to find its own way through the galaxy.

Big babies

A starburst is blazing in the Large Magellanic Cloud, a dwarf galaxy that orbits the Milky Way about 180,000 light years away. It's located in the nebula 30 Doradus, which is is about 50 light years across and visible through binoculars as a smudge in the skies of the southern hemisphere. It is forming stars in a flat-out sprint, seeding new ones more than 10,000 times as fast as the Milky Way does in our part of space.

Any residents would see blue stars brighter than the full moon at night, shining through the spidery veins of dust and gas that hang through the nebula like cobwebs.

Among its many attractions, 30 Doradus boasts the biggest stars we know of in the universe – so big, indeed, that they should be impossible. Theorists have long thought that there ought to be a limit to how heavy a star can grow. Bigger stars burn faster and shine much brighter than small ones, and at some point their sheer radiance should eject their own outer layers. Theorists thought this outward force would prevent stars growing to more than about 120 times the mass of the sun.

But some of the heaviest stars in 30 Doradus seem to weigh in at 180, 195, and an obscene 325 suns. To explain those measurements, theorists have had to turn to ever more complex computer simulations – and while they are making progress on an answer, they still disagree about the specifics.

Such a brief, violent generation of stars changes the surrounding environment with blasts of ultraviolet radiation and charged particles. After a few million years, these stars and the supernovae that follow will have blown away all the gas around them.

The starburst plays the role of a wildfire on a plain, burning fast and bright. Although both are destructive, consuming all in their wake, they scatter nutrients: organic-rich soil on the plain, and heavy elements forged by stars in the case of the starburst, which will seed future planets. When it's over, the area lies fallow until something new can grow.

We think 30 Doradus is already past its peak. In a few million years, without gas to make new stars, the show will be over – except in surrounding regions, which might be triggered into forming stars of their own.

The cool ones

As if space weren't lonely enough, pity the brown dwarf. Compared with their stellar siblings, these astronomical objects are something of a failure. And while they have much in common with planets, they don't seem to fit in there either. This awkward status as cosmic in-betweener means brown dwarfs are often overshadowed by their flashier counterparts, such as alien worlds or fiery supernovae. Yet not fitting in is precisely what makes brown dwarfs far more interesting and useful than we once thought.

The existence of brown dwarfs was first suggested in 1962 by Shiv Kumar at the NASA Goddard Institute for Space Studies in New York City, who had wondered how small a star could be. Below a certain size, Kumar calculated, you would end up with objects with too little mass to sustain hydrogen fusion.

Kumar called these hypothetical objects black dwarfs, but the name proved problematic. In the 1970s, astronomer Jill Tarter pointed out that the term also referred to a dark, cooling star near the end of its life. Various other names had been proposed, such as planetar, still-born star and substar, but Tarter argued for brown dwarf. She knew they couldn't actually be brown, but she felt labelling them with a composite colour was appropriate since their actual colour was going to be difficult to observe due to their feeble radiation. (Zoom past in a spaceship and you may well fail to see one because it would produce so little visible light. Peer closer, and you might see a faint glow in regions where it is still hot enough to produce light – but it might be more of a very dark orange.)

None was spotted for 20 years, but finally in 1995, Gliese 229b popped into view. About 19 light years away, this brown dwarf has a mass between 20 and 50 times that of Jupiter, and a relatively cool surface temperature of 680°C. Since then we have found thousands of brown dwarfs with puzzling traits, prompting a debate about how to classify them.

Stars or planets?

Since humans first looked to the heavens, there has always been a separation between stars and planets. Brown dwarfs challenge these ideas. They are born from the collapse of a gas cloud, just like stars, so share some features with their stellar relations. They have magnetic spots like stars, and some even emit radio emissions like pulsars (see Chapter 6). Many are also big enough to spark a brief burst of nuclear fusion at the start of their lives, burning up a small stock of deuterium (and larger ones probably burn a little lithium too). Along with the heat generated by gravitational collapse as they form, this makes young brown dwarfs hot enough to shine faintly in visible light. Gradually, they cool down. About seven light years away, WISE J085510.83-071442.5 has reached temperatures well below zero °C.

Such cold objects, without sustained nuclear fusion, seem very unlike ordinary stars. So should we think of them more like planets? Brown dwarfs are much more massive than most planets – between 13 and 70 times the mass of Jupiter. Only 3 to 4 per cent of known exoplanets are so hefty. But there is certainly some overlap, and in actual size most brown dwarfs are not far off the diameter of Jupiter. They have similar atmospheres to gas giant planets, too: toxic brews of carbon monoxide, hydrogen sulphide and water, or methane and ammonia.

And they have weather. It was always suspected that brown dwarfs had clouds, because their internal heat would prompt gases to rise and then condense, as happens in the atmospheres of planets in the far reaches of our solar system. But recently we have been able to watch this weather change over time. By training their telescopes on a target for months at a time, astronomers can follow changes in infrared emission caused by huge storms.

We know from studying the chemical composition of stars that the atmospheres of hotter brown dwarfs contain gaseous iron and silicate, which would eventually condense as it rises and cools. Imagine rain drops of molten iron, with swirling clouds made of hot grains of sand that gradually fall as silicate snow. Meanwhile the coolest have weather patterns we associate with our own planet – a few may even have clouds made of water vapour.

Perhaps the most tantalizing revelation is that brown dwarfs can be accompanied by planets. In 2013 a team of astronomers found a gas giant planet orbiting a brown dwarf. Many of the future planets we find are likely to be small and rocky, as young brown dwarfs have less material surrounding them than more massive stars. Life, then, could exist on a world orbiting a brown dwarf.

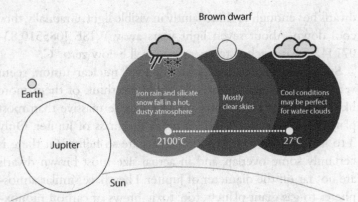

FIGURE 5.3 Brown dwarfs share many traits with gas giants such as Jupiter, overlapping in their ranges of size and temperature. This means their atmospheres could provide clues to exoplanet weather.

Given that brown dwarfs have now finally come of age, perhaps we should stop trying to pigeonhole them as planets or stars. It's time to put them in a class of their own.

Serial exploder

Even in a universe full of weird objects, Eta Carinae is an oddball. In 1843, it briefly became the second brightest star in the night sky, erupting like a supernova yet somehow managing to survive. And recently we have found evidence of more such outbursts, occurring around 1550 and 1250 CE.

Located 7500 light years away in the constellation Carina, the object is actually two massive stars spinning around one another in a tight 5.5-year orbit, with a combined brightness more than 5 million times that of our sun.

The smaller star is estimated to be between 30 and 50 times the sun's mass, while the larger is a behemoth of between 100 and 150 solar masses. This star is tearing itself apart. Its photons blast outward with so much pressure that they are carrying away the star's outer layers.

In 2016 that shed skin helped researchers to reconstruct Eta Carinae's violent past. Using photos from the Hubble Space Telescope taken two years apart, Megan Kiminki at the University of Arizona in Tucson and her colleagues made a movie tracing more than 800 gas blobs from Eta Carinae back through time.

Some gas clouds are moving up to 3 million kilometres per hour. But crucially, the filaments don't seem to be speeding up or slowing down, letting Kiminki's team estimate when they were emitted from the system. They suggest that Eta Carinae had a major eruption around the year 1250 CE, and a slightly less dramatic one near 1550 CE. Several very distant clumps could have come from events in 1045 and 900 CE, though these might have been shot out at high speeds from the thirteenth-century event.

Nobody knows why Eta Carinae goes through these periodic outbursts. Some suspect the companion to the larger star might occasionally interact with its outer layers, dumping in fresh nuclear material and causing flare-ups. Perhaps we'll be lucky enough to see this volatile giant explode again soon.

Do stars collide?

Even though space is vast and almost empty, there must be some stellar collisions. We may even have seen one. In February 2002, a previously undistinguished star called V838 Monocerotis, about 20,000 light years away, briefly achieved a luminosity a million times that of our sun. The following month it happened again. And again in April. It was first assumed to be a nova − a white dwarf that pulls gas off a companion until it triggers a thermonu-clear explosion on its surface. But novae don't happen three times in quick succession and then go quiet. One hypothesis is that this was a mergeburst: the scream of two stars colliding. But it could also have been a rarely seen thermonuclear flare-up near the end of a giant star's life – or maybe a star swallowing giant planets. In any case, the result is a strange and beautiful object, with the triple burst of light being reflected off nearby dust to surround the object with rapidly changing shells of light.

What's the biggest known star?

The leading contender at the moment is UY Scuti, a red supergiant about 10,000 light years away. It is estimated to be 1700 times the diameter of the sun. If it were to suddenly replace the sun, Jupiter would be orbiting deep inside it. Not that Jupiter would be destroyed – the outer parts of a red supergiant are so tenuous that they would pass for a perfect vacuum here on Earth.

6
The afterlife of stars

When a giant star explodes, its story is not over. Within a glowing cloud of debris, the star's heart may be preserved as a furiously spinning, ultramagnetic, overgrown atomic nucleus. Or it could be even more remarkably transformed, into a piece of disembodied gravity, a breach in spacetime.

The case of Kepler's supernova

The facts of the case are as follows. On the night of 9 October 1604, Europe turned to the south-western sky, where Jupiter, Saturn and Mars were slated to assemble in Sagittarius. Some believed it would herald a radical transformation of the world.

The conjunction went as scheduled. But something else stole the spotlight in the nearby constellation Ophiuchus. A new star appeared, growing brighter and brighter for 20 days, becoming more luminous than any planet, lingering, and finally fading over the next year. It was the last great Milky Way supernova recorded by observers with their naked eye.

'We can be sure of only one thing,' wrote Johannes Kepler, who compiled detailed records of the event. 'Either the star signifies nothing at all for mankind or it signifies something of such exalted importance that it is beyond the grasp and understanding of any man.' Today's astronomers – if they're feeling grandiose, at least – might lean towards that second option.

The still-expanding stellar shrapnel is called Kepler's Supernova Remnant. Studying it is a bit like cosmic blood-spatter analysis. In hindsight, astronomers classify what happened back in 1604 as a type Ia supernova: the kind that modern cosmology uses as a measuring stick to gauge the size and history of the universe.

Despite how much we depend on them, what causes type Ia supernovae in general is not certain. In one model, mass from a neighbouring red giant star falls on a dense, hot white dwarf core, which then obliterates itself in a thermonuclear explosion. The alternative is that type Ia supernovae occur when two white dwarfs merge, wiping each other out.

Can Kepler's supernova help to clear this up? The scene does contain what many think is a key clue: gas ejected from the supernova seems to be ploughing into other gas that was ejected earlier from the system. That hints at the involvement of a red giant star, which would have been expelling some of its atmosphere to space, rather than colliding white dwarfs.

But searches for that second star have drawn a blank. This could mean that there was a second star next to a white dwarf, but that the second star transformed into a white dwarf, too, shortly before the pair annihilated each other. Or maybe the second star is still hiding there, but disguised or disfigured by the blast – and is now fainter or otherwise unrecognizable.

Astronomers are holding out hope that deeper searches for that second star could still find it, or that spectral studies of the remnant might provide new evidence dating back to the time of the explosion. Until then, a cold case smoulders white-hot in the south-western sky.

Stellar makeover

Neutron stars are the cores of massive stars that exploded as supernovae. Whereas type 1a supernovae are powered by nuclear reactions, other supernovae are powered by gravity. When large stellar cores finally run out of fuel, with no new heat to support them, they collapse under their own gravity, squeezing down to extreme density – until a new force finds the power to save them. The strong nuclear force, which normally holds protons and neutrons together in atomic nuclei, becomes repulsive when nuclear matter is compressed enough.

Gravity and the strong force reach an impasse when the core has shrunk to about 10 to 15 kilometres across. Under these conditions, it's thought that most of the protons and electrons combine to make neutrons. These particles are packed together so tightly that a teaspoonful of the material would weigh billions of tonnes. It is thought to be superfluid, flowing without friction, and threaded with magnetic vortices.

As if this stuff were not strange enough, some physicists have speculated that in especially massive neutron stars, the enormous pressure could cause the neutrons to break down, freeing their individual quarks. Others suggest that the particles will form a Bose-Einstein condensate (BEC), a quantum state in which the neutrons' individual identities blur and they behave collectively as a single particle.

These exotic-matter theories received a blow from studies of a neutron star called EXO 0748-676. Its mass has been measured at about twice that of the sun. Most models of quark stars and BEC-containing neutron stars predict they would collapse into a black hole before reaching such a high mass.

But the case is not quite settled – and even if the basic material is neutrons after all, it could hold some surprises. In 2014 Charles Horowitz at Indiana University Bloomington and his colleagues simulated a tiny box of neutron star material smaller than a single atom, containing tens of thousands of neutrons and protons. The strong nuclear force and the electrostatic force fight over the packed protons and neutrons and drive them into strange shapes resembling waffle-like grids. The waffle structures have features just a shade bigger than an atomic nucleus.

Starquake

The crust of neutron stars is not so highly compressed. It may be more like familiar solid matter, with nuclei and electrons. This stuff is still ultra-strong, but it can be torn apart by the magnetic field of certain neutron stars known as magnetars. Their fields are so strong that if one passed halfway between

Earth and the moon, it would wipe the data off every swipe card on Earth. It's thought that as the fields inside a magnetar twist around they can rip the crust open, releasing a fireball of particles and radiation that astronomers observe as a bright flash of high-energy photons. This is a starquake.

Back in 2006, astronomers used a particularly powerful starquake to measure the thickness of a neutron star crust. It was picked up by NASA's Rossi X-ray Timing Explorer in December 2004, from a star called SGR 1806-20. A team led by Tod Strohmayer of NASA's Goddard Space Flight Center found that the quake set the neutron star ringing, with oscillations of various frequencies appearing in the X-ray spectrum. The team think that some waves were ringing through the crust vertically, letting them calculate its thickness: about 1.5 kilometres.

Magnetars may also explain some super-bright supernovae, as their whirling magnetic fields could pump extra energy into the cloud of debris thrown out by an initial supernova explosion that formed the magnetar.

Cosmic clocks

Night in, night out, rhythmic radio signals reach Earth. The slowest of them sounds like a nail being hammered into wood, or a shoe being slapped against a post to rid it of mud. Others are more like a stuttering motor stopped at a traffic signal. Some make almost continuous tones, ripe to be combined into cosmic mood music. Always the same signature tunes, always from the same points in the sky. Small wonder that when astronomers heard the first of them in 1967, they briefly wondered whether it was a message from aliens.

What they had actually found was a pulsar – a type of neutron star that sends out regular radio signals. For a neutron star to be a pulsar, its magnetic axis must be at an angle to its rotational axis. Then powerful jets of radiation erupting from the star's magnetic poles will sweep round as the star rotates, rather like the beam of a lighthouse. These jets are what regularly buzz our telescopes – although we still don't know exactly how they are formed.

One recent suggestion, by John Singleton and Andrea Schmidt of the Los Alamos National Laboratory in New Mexico, is that it's akin to the sonic boom produced by supersonic aircraft as they accelerate past the speed of sound. Relativity does not forbid the magnetic fields at the surface of a pulsar rotating faster than the speed of light, says Singleton. As they do so, his team suggests, particles of opposite charge are pushed to either side of the pulsar, where they emit radiation. The pattern of radiation is then sharpened by the superluminal boom of the magnetic field into a sharply defined pulse that is emitted into space.

Whirls and waves

In 1974, astronomers Russell Hulse and Joseph Taylor discovered one pulsar circling particularly tightly around a companion, completing one orbit every eight hours. They saw the distance between the two bodies steadily diminish as they spiralled in towards each other, at exactly the rate calculated if they were losing energy by radiating gravitational waves. This was our first evidence for the travelling distortions in spacetime predicted by Einstein's general theory of relativity.

The first pulsars to be discovered spun in a comparatively leisurely fashion, taking several seconds to complete one rotation. In 1982, however, a group led by the late Donald Backer upped the ante with one that whirls around a breathtaking 642 times a second, put in its dizzy spin by matter siphoned from a companion star. We've since found more of these millisecond pulsars. Their pulses are so fast and regular that they make fantastic clocks, and some astronomers are now monitoring them for any slight changes in timing caused by passing gravitational waves.

FIGURE 6.1 Jocelyn Bell Burnell discovered a new radio signal in 1967. It blared out so regularly she jokingly called it LGM-1, for little green men.

Interview: Putting her finger on the pulsar

Fifty years ago, Jocelyn Bell Burnell discovered a mysterious, pulsing radio signal – and the downsides of being a young woman in science. New Scientist *interviewed her in 2017.*

In a way, it was the second signal that was the big one. The first signal I saw could have been a mistake. The second one meant this was something real. It took a while to realize what we had found: the very first pulsars, a new type of star. We're still working out the true significance of the discovery today.

It was 1967, and we were looking for **quasars** using a radio telescope designed by Tony Hewish, my supervisor at the University of Cambridge. Back then, we knew only that quasars were very distant objects, with radio signals that grew strong and weak in an irregular way. But this new signal was strong, not weak, and came in absolutely regular short bursts.

It didn't look like interference, either, although that was often a problem for us. Our telescope was a tangle of 2048 radio antennas covering about 4 acres just outside the city. You pick up a lot of interference with such a vast collecting area. Once somebody mistakenly allocated our observation frequency to the local police.

The first unexpected signal was jammed into a quarter-inch of the chart recorder – just a pen moving mechanically over paper – which I'd set to run slowly for the longer quasar signals. So I ran the paper faster at the time of day the signal was appearing, to spread it out, a bit like a photographic enlargement. But nothing came. The signal had disappeared.

One of the first questions my colleagues asked was whether I had wired the telescope up wrong. I was used to that. For one thing, I was a junior doctoral student. For another, I was a woman. It had been worse in Glasgow, where I'd done my undergraduate degree. There, whenever a woman entered the lecture theatre, all the guys whistled, stamped, banged the desks and catcalled. Cambridge was more genteel, but also more snooty. I felt like an impostor there, as a girl from the provinces, from Northern Ireland. I was convinced that someone would find me out and then throw me out. I worked as hard as I could, so I'd have a clear conscience when that happened.

After a month or so, the signal reappeared. I immediately phoned Tony. If it was a signal, he said, it must be of human origin because it was so regular, pulsing once every 1.3 seconds, like a metronome beat. But I knew it couldn't be. Stars rise and set 4 minutes earlier each day as Earth orbits the sun. It had been early August when I first observed the signal. Now it was November, and the signal had kept pace with the stars. If it was something artificial, like radio interference from someone driving around in a car with a badly suppressed alternator, they would have had to religiously get 4 minutes earlier every day.

It was an anxious moment when Tony came out to the observatory the next day to look over my shoulder, but, sure enough, the signal came. That was when we had to start thinking about what on earth – or off it – it might be. I called it LGM-1, for 'little green men', as a joke. But if it was a communication from an alien intelligence, they were using a bloody stupid technique. For one thing, the signal was amplitude-modulated. There are many ways that natural phenomena can modulate a signal's amplitude. If you want to signal across light years of space, you wouldn't use AM, you'd use FM – it makes for a more obviously artificial signal.

We managed to estimate the source's distance. It was about 200 light years away: within our galaxy, but way beyond the distance our TV and radio signals had travelled into space since they'd started a couple of decades earlier. It really would have been a curious set of little green men signalling to our inconspicuous solar system.

That was when we found a different signal and, a few weeks later, a third and a fourth, each with its own periodicity. That demolished the little-green-men hypothesis, unless lots of aliens were signalling to us from opposite sides of the universe. Instead, it must be some new kind of star. We didn't know that when we published our paper in *Nature* – 'Observation of a Rapidly Pulsating Radio Source' – in February 1968. Of course, the media only latched on to one line in the paper saying we had briefly considered the signals might have originated on an alien planet.

I published under the name S.J. Bell, and at first the press didn't realize I was a woman, let alone a young one. When they found out, I suddenly had reporters on the phone asking if I was brunette or blonde – no other colours were allowed, apparently – and what my vital statistics were, which I didn't know. I was asked questions like how tall I was, and was that taller than Princess Margaret or not quite so tall? And then photographers were asking me if I could please undo the top buttons of my blouse. I have a sharp tongue and I would have loved to use it, but

I wasn't in a position to do so. The lab needed the publicity and I needed good references for my next job.

It was a similar story in 1974 when Tony Hewish was awarded a share of the Nobel prize for the pulsar discovery and I wasn't. I said at the time it was only right, because he was my supervisor, but I didn't entirely believe it. I don't think the snub was because I was a woman. It was because I had been a student. In those days, students just weren't recognized. That's changed for the better since.

I got married soon after the pulsar discovery and moved away from radio astronomy, following my husband's relocations for his job. I've had a varied career since: I've done astronomy in most bits of the spectrum, and been a lecturer, a researcher, a tutor and a manager. But I still feel a bit proprietary about pulsars, so I've kept a friendly eye on them.

Aftershock

Long after a supernova itself has faded, the gassy remnants of the explosion continue to expand – sometimes forming beautiful nebulae like the Crab (Figure 6.2). These wispy clouds turn out to have a more violent side, and regularly kill human beings.

Cosmic rays are charged particles arriving at Earth from space. Nearly all of them are protons, and some have been accelerated to speeds higher than any achieved by a particle accelerator on Earth. Although we have known about cosmic rays since 1912, their origins have remained a puzzle.

Physicists suspected the main source might be **supernova remnants**. The material blown out from a supernova moves so quickly that it creates a shockwave, where tangled magnetic fields converge.

Because protons are charged, they can get caught in these fields which carry them back and forth across the shock many times, like a tennis ball bouncing back and forth, gaining energy each time.

But this was hard to prove. Interstellar magnetic fields can deflect cosmic rays on their way to our detectors, so by the time they reach Earth their directions are scrambled, making it hard to determine their origin. Another approach to the problem was needed – and gamma rays provided it. When a high-energy

FIGURE 6.2 The Crab nebula, remnant of a supernova seen by Chinese astronomers in the year 1054. It is about 6000 light years away, about 10 light years across and holds a bright, rapidly spinning pulsar.

proton collides with a low-energy protons, it can create gamma rays with a characteristic minimum energy. These are uncharged and so travel in straight lines, unaffected by magnetic fields.

Using the Fermi Gamma Ray Space Telescope, Stefan Funk of the SLAC National Accelerator Laboratory in Menlo Park, California, and colleagues observed two supernova remnants. They saw lots of gamma rays above that characteristic energy – and almost none with lower energies, confirming that these remnants are active particle accelerators.

It doesn't explain the origin of all cosmic rays. Some are muons or positrons instead of protons, and some, the ultra-high energy cosmic rays, are probably from outside our galaxy. But it looks as though the bulk of cosmic rays, which provide a large part of our background radiation dose on Earth, do come from supernova remnants.

Ultimate implosion

Sometimes even the strong nuclear force is not strong enough. Stars of more than about 20 solar masses have cores so large that when they run out of fuel and start to collapse, no known force can oppose the inward rush. Gravity is the inevitable victor, dragging material inwards, to form what we call a black hole.

If you use Einstein's general theory of relativity to describe the gravitational field of these things, you find that right at the centre the curvature of space time becomes infinite – forming a feature called a singularity, a hole in the fabric of space time. An even stranger feature is an invisible spherical surface, known as the event horizon, surrounding the singularity. Nothing can escape once it has passed the event horizon.

Well, almost nothing. Stephen Hawking showed that black holes may not be entirely black: a quantum froth of particles and antiparticles popping into existence near the horizon should produce a form of radiation now known as Hawking radiation. In the unimaginably distant future it could mean that black holes eventually lose all their energy and evaporate away.

Although no black holes have been seen directly, there is overwhelming evidence that they exist. They are normally detected through the effect they have on nearby astrophysical bodies such as stars or gas. In 1972, an object about 6000 light years away called Cygnus X-1 was identified as a likely black hole. It orbits a blue supergiant, and as gas from this companion star spirals into the hole it heats up and emits intense X-ray radiation. With a mass about 15 times the sun's, it is far too massive to be a neutron star – so a black hole seems the most likely option. Cygnus X-1 was the first of many such X-ray binaries to be identified as black hole candidates.

It's thought that a black hole birth would normally be heralded by a brilliant supernova, but for stars at the lower end of the mass scale the new black hole might swallow most of the material around it, snuffing out the explosion. We may finally have seen a black hole being born in one of these failed supernovae.

In 2016, a team led by Christopher Kochanek at Ohio State University in Columbus reported glimpsing something special in data from the Hubble Space Telescope. The red supergiant star N6946-BH1, which is about 20 million light years from Earth, was first observed in 2004. For some months in 2009, the star briefly flared up, then steadily faded away. New Hubble images show that it has disappeared in visible wavelengths.

These observations mesh with what theory predicts might happen when a star that size crumples into a black hole. First, the star spews out so many neutrinos that it loses mass. With less mass, the star lacks enough gravity to hold on to a cloud of hydrogen ions loosely bound around it. As this cloud of ions floats away, it cools off, allowing the detached electrons to reattach to the hydrogen. This causes the bright flare. When it fades only the black hole remains.

The true nature of black holes is still unknown. Some theorists speculate that if you fall into a hole, just beneath the horizon a 'firewall' will destroy you; others suggest that you might fall through a wormhole and into another universe. What happens near the singularity is a true mystery. Neither relativity nor quantum theory can answer that question, and physicists are struggling to find a unified theory of quantum gravity that can.

Black hole sun

If planets orbit a black hole – as in the film *Interstellar* (2014) – you might expect them to be among the most inhospitable places in the universe, cold and dead. But it is possible that they could sustain life, thanks to a bizarre reversal of the thermodynamics experienced on Earth.

According to the second law of thermodynamics, life requires a temperature difference to provide a source of useable energy. Life on Earth exploits the difference between the sun and the cold vacuum of space. So what if you flip the temperatures around, with a cold sun and a hot sky?

Some black holes are among the brightest objects in the universe, shining brilliantly not only in visible light but often also in radio, infrared, UV, X-rays and gamma rays. That is because gas and other matter falling in is superheated and glows as it accretes. But a satiated black hole effectively has zero temperature, meaning it could potentially act as a cold sun, according to Tomáš Opatrný of Palacký University in Olomouc, Czech Republic.

The rest of the sky has a temperature of 2.7 kelvin (about −270 °C), thanks to the cosmic microwave background (CMB), the heat left over from the explosion of the big bang. In 2015 Opatrný's team calculated that an Earth-sized planet orbiting a black hole that appeared a similar size to our sun in the sky could extract around 900 watts of useful power from this temperature difference – perhaps enough for life to exist, but hardly impressive.

Wondering if any more power might be available, the team turned to the film *Interstellar*, in which a world called Miller's planet orbits very close to a massive, spinning black hole called Gargantua. General relativity means the black hole's gravitational pull slows time on the planet so that one hour is equal to seven years off-world, a factor of around 60,000.

The energy of light is proportional to its frequency. This means that when light from the CMB hits Miller's planet, and its frequency is increased by this time dilation, its energy increases. With a time-dilation factor of around 60,000, Miller's planet would be heated to nearly 900 °C.

In the film, the planet is swept by huge tidal waves of water, but Opatrný says his calculations mean molten aluminium would be more likely. Conditions would be cooler if the planet were slightly further out from the black hole, lessening the effects of time dilation and making it more hospitable to life.

As other researchers point out, in practice this may be an unlikely setup. Even if planets can form in such an orbit, there is liable to be some matter falling into the hole and emitting heat.

What is the nearest black hole to Earth?

The nearest known black hole candidate is in the X-ray binary V616 Monocerotis, where observations suggest that a black hole of seven solar masses orbits an orange star. It is about 3000 light years away. But there are probably millions of black holes in our galaxy, still undetected, so some of them will be much closer.

7
A trillion planets

Our galaxy is teeming with planets. The thousands we know of are strikingly diverse, from hot and huge, to cool, rocky and perhaps habitable, via many shades of downright peculiar. What awaits among the many billions that are yet to be detected – and can we hope to find alien life out there?

The exoplanet zoo

Before 1995, the planets in our solar system were the only ones that we knew for certain existed. Then PhD student Didier Queloz at the University of Geneva discovered the first planet circling an alien star, and soon the floodgates of discovery opened wide. Now with more than 3500 confirmed exoplanets and counting we've discovered an enormous variety of worlds.

The search for life beyond Earth has always been a top priority for humankind. Life as we know it requires light, water, mild temperatures and moderate gravity, so rocky, Earth-like worlds are especially appealing. Particularly if those exoplanets are in the Goldilocks zone, where the temperature is just right to keep water liquid.

The TRAPPIST-1 planetary system is a shining example. Just 40 light years away, seven temperate Earth-size planets orbit within a stone's throw of each other, each offering its own chance for atmospheres, oceans and life to emerge. The discovery suggests such Russian-doll systems of nested small worlds may be common, and perhaps the best places in our galaxy to look for life. Their orbits are tightly spaced, and through the exchange of gravitational tugs they have settled into harmonies. For every eight times the innermost planet circles its star, the second planet orbits five times, the third planet orbits three times, and the fourth planet orbits twice. That kind of compact gravitational clockwork might even facilitate the spread of life between worlds.

A newcomer called LHS 1140b is already being hailed as the best place to look for signs of life beyond the solar system. It is a super-Earth, about 1.4 times the diameter of our planet and 6.6 times as massive. Despite a close orbit around its star – it circles once every 25 days – the star's dim nature means it receives about half as much light as Earth does, so is probably cool enough for liquid water. The star is a red dwarf, which are often prone to flares that could damage life on nearby planets, but LHS 1140b is unusually calm, so its planet is at lower risk. Only 40 light years away, LHS 1140b is one of the closer potentially habitable exoplanets we have found. As it is so close and passes between its star and us, it presents an excellent opportunity to look for an atmosphere distorting the star's light as the planet passes by – and to seek out signs of life in that atmosphere.

The nearest rocky exoplanet we have spotted is Proxima b, only 4.2 light years from Earth. Its size hasn't been pinned down, but it is probably just a bit bigger than Earth. It orbits close to the red dwarf Proxima Centauri, experiencing a lot of X-ray radiation and a strong **stellar wind**, which may not be ideal for life. The upside is that Proxima b is right next door, which could let us look for an atmosphere and signs of life there, and perhaps one day send a probe.

Alien hellscapes

Most exoplanets look hellishly inhospitable. The hottest yet found is KELT-9b. Twice the size of Jupiter, it orbits close to a hot star. The planet's dayside temperature reaches more than 4300°C, as hot as an orange star. The intense light and heat are enough to evaporate the gas giant's atmosphere at a rate of up to 10 million tonnes per second, which may render the planet a naked core by the time its star expands and envelops it.

If extreme heat isn't bad enough, how about volcanoes and lightning? Kepler-10b, the first confirmed rocky world outside our solar system, is so close to its star that its surface may be made up entirely of volcanoes. Volcanic dust often leads to lightning bursts, and based on data from eruptions on Earth,

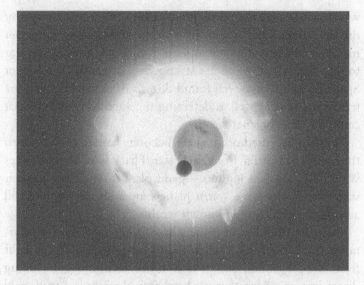

FIGURE 7.1 Some exoplanets have been seen transiting binary stars (artist's impression).

one model shows that Kepler-10b may see a trillion lightning flashes in an hour.

Early weather reports from one planet 1000 light years from Earth point to clouds of ruby and sapphire. The brightness of a planet called HAT-P-7b changes over time, indicating that the brightest areas on the planet move around with time, which may be due to changes in cloud coverage around the world. Given the high temperature of the planet, the clouds may be made of corundum, the same mineral that forms sapphires and rubies on Earth.

Planet collecting

Most exoplanets have been spotted by one of two methods. The planet's gravity can cause its host star to wobble slightly, which is revealed as a subtle doppler shift in the wavelength of the star's light. Or, if the orbit is in the right plane, the planet may pass in front of its star as seen from Earth, causing the star to dim periodically – an event known as a **transit**. The Kepler Space Telescope alone has found thousands of planets using the transit method, as well as detecting the variations in light that can indicate exoplanet weather.

With either method, planets become harder to see when they are small or far from their star. That's why the first ones we found were hot Jupiters – giant planets very close to their stars – and why large warm planets are still over-represented among the catalogue of known worlds.

A smaller number of planets have been spied out by gravitational microlensing – when their gravity magnifies the light from a background star – and other techniques, including direct imaging. One that we have seen directly is a truly cold and lonely planet, 2MASS J2126, which orbits a trillion kilometres away from its dim brown-dwarf star. That is 6900 times further from its star than Earth is from the sun. Images have also captured the motion of four planets in the system HR 8799.

Along with hot Jupiters, other new classes of planet have been loosely defined, such as super-Earths (large rocky planets) and mini-Neptunes, the smallest class of gas giants. Still, not everything fits in. Kepler-10c is so unusual that it has been placed in a brand new class of exoplanet. It orbits a star that is about 560 light years away from us, and has a radius slightly more than double that of Earth's – a size that led astronomers to assume it was a mini-Neptune. But in 2014 we found that it is actually 17 times as heavy as Earth, which given its diameter means it must be an incredibly dense, solid world the like of which has never been seen before. Planetary formation models had not predicted such a thing. Kepler-10c is the first mega-Earth.

The old ones

Already ancient when the Earth was born, five small planets orbiting an orange dwarf star are about 80 per cent as old as the universe itself.

NASA's Kepler Space Telescope spotted the planets around a star called Kepler 444, which is 117 light years away and a little smaller than the sun. Orange dwarfs such as this are considered good candidates for hosting alien life because they can stay stable for up to 30 billion years, compared with the sun's 10 billion years. As far as we know life begins by chance, so older planets would have had more time to allow life to get going and evolve.

Kepler 444's planets range from 0.4 to 0.74 times Earth's radius. A technique called astroseismology revealed the age of the star at about 11 billion years. For context, the universe is 13.8 billion years old. That makes this the oldest known system of terrestrial planets in the galaxy.

While Kepler 444's planets are much too hot for life, their existence suggests that there might be cooler old worlds elsewhere.

Strange new worlds

Exoplanets, we have learned, are happy to bask in the simultaneous light of four suns, wander the galaxy as starless outcasts from their home solar systems or orbit whipper-snapper stars barely 1 million years old. Naive theories of what extrasolar worlds should be like have been consistently outshone by the data, and there are probably plenty more surprises in store.

In an attempt to steal a march on nature, researchers are now busy imagining weird new types of exoplanet that might turn up in future. Far from being a parlour game for bored astronomers, understanding nature's ability to produce planetary bodies will be crucial to learning how our solar system compares to its galactic counterparts. Many of these proposed exoplanets challenge our understanding of planet formation, and could play havoc with our admittedly arbitrary criteria for determining what constitutes a planet in the first place. Furthermore, exotic new planetary types should expand our inevitably Earth-centric ideas about where habitable planets might form, aiding in the search for alien life.

Binary worlds

In our solar system, large planetary bodies are located far apart, orbited by moons of much smaller size. We think this familiar configuration emerges when bits of dust clump together in a protoplanetary disc encircling a young star, evolving into rocky hunks that hoover up any material in their orbital paths. Moons can then be crafted from leftover detritus orbiting the planet, or be hauled in during the chaotic pinballing of objects thought to happen in developing solar systems.

There is a third option, however. Widely accepted models suggest our own moon formed when a Mars-like body smacked into the primordial Earth, gouging out material that coalesced into the satellite we know today. But if those two bodies had undergone a less spectacular collision, they could have ended up in a stable partnership as a binary planet.

Finding binary planets could shed light on the rambunctious childhood years of fledgling solar systems. It would also prove that collisions of the kind that created our moon can be considered a viable route to planethood, and not just a way to form satellite hangers-on. Fortunately for astronomers, binary exoplanets should cast distinctive double shadows as they cross and partially **eclipse** the shining faces of their stars – transit signals readily detectable by NASA's Kepler Telescope and other observatories designed to look for new worlds.

Undoubtedly the most intriguing configuration would be two Earthlike worlds locked in a binary orbit. Imagine if Earth had a habitable twin in our sky, and life, or even a spacefaring civilization, arose there in parallel to our own.

Party planets

Although the worlds in our solar system stick standoffishly to their own orbital lanes, they do tolerate company beyond their faithful moons. Asteroids dubbed Trojans, for instance, hang out at Lagrangian points, sweet spots where the gravitational force of a planet and its star combine to keep any inhabitants orbiting in sync with the planet. Jupiter shepherds an army of Trojans around the sun, and Earth has a Trojan of its own, a rock called 2010 TK7.

In theory there is no reason why planet-sized objects couldn't arrange themselves in a 'party orbit', all at roughly similar distances from their host star. Arrangements of this sort can be stable for billions of years – so long as there are no gravitationally perturbing worlds on either side of the crowded orbital band to disrupt its delicate choreography. What is less clear is how they could form in this configuration.

The existence of co-orbital planets would upend the prevailing doctrine that planets must keep their orbital backyards free of other large bodies – something that ousted Pluto from the full-planet club in 2006. These worlds could even cross-pollinate, thanks to meteorite impacts blasting out rocks harbouring hardy bits of genetic material.

Egg worlds

The gaseous giant WASP-12b orbits its star at such scorchingly close quarters that the strong stellar gravitational pull has warped it into a bulging oval. Prabal Saxena and his colleagues at George Mason University in Fairfax, Virginia, decided to explore how this tidal distortion might affect a rocky world such as Earth. They calculated that an exoplanet of this kind could stretch to be a fifth wider at its equator than from pole-to-pole before being torn apart.

The discovery of such rugby ball–shaped worlds could be a boon for planetary science. The way a planet responds to a star's gravitational pressure would provide an entirely new way of learning about its insides – for example showing whether it is mostly solid or gaseous. As a bonus, the atmospheres of ovoid worlds would experience different levels of gravity in different places, possibly making for intriguingly unpredictable climates.

Chthonian planets

As solar systems evolve, planets may migrate inwards or outwards. Nudged towards a stellar furnace, some gassy planets may have their atmospheres stripped away by the heat and stellar winds, eventually leaving nothing but their rocky cores. This exposure of the planet's hidden depths inspired the name chthonian, a reference to the deities of the Greek underworld. If a cold mini-Neptune migrates towards its star's temperate, habitable zone, the extra heat could not only blow off the atmosphere but also melt exposed surface layers rich in water ice. These planets could transform into ocean-covered worlds with life-friendly air.

Binary worlds

Two planets of equal size, orbiting each other as they journey around a shared star

Party planets

Multiple worlds occupying the same orbit, violating the conventional definition of a planet

Egg worlds

Rocky planets squeezed into highly eccentric shapes as a result of their star's strong gravitational pull

Chthonian planets

Habitable evaporated cores (HECs)

Gassy worlds born in a distant orbit that migrate towards their star. The sudden heat would evaporate their atmospheres, thawing their frozen cores into oceans capable of sustaining life

Exposed cores

Gaseous giants pulled towards their stars, where their thick atmospheres are boiled away to expose their hidden, rocky cores

Corkscrew planets

Planets tracing a corkscrew orbit around the axis between two stars

FIGURE 7.2 Yet-to-be detected exoplanets could have shapes and orbits unlike anything yet seen.

Corkscrew planets

Mind-bendingly, some worlds could exist in a sort of orbital limbo, spiralling about an axis between two stars in a binary system, pulled hither and thither by their competing gravities. The brainchild of theoretical physicist Eugene Oks of Auburn University in Alabama, these whirligig worlds would follow a completely new type of stable, albeit speculative, planetary orbit.

Oks ran the numbers for a corkscrew planet pulled between the orange and red dwarf stars comprising Kepler-16, a binary system 200 light years from Earth. The planet would complete a manic loop-the-loop in its cone-shaped orbit in under an Earth week.

Any life that managed to survive on such a permalit planet, where seasons change in the span of days, would experience one of the weirdest night skies in the cosmos. Upon reaching one end of the corkscrew orbit and heading back towards the other star, the closest sun would seem to suddenly reverse its direction in the sky.

Seeking ET

Stand by a payphone and wait for it to ring: that sums up most of our attempts to search for extraterrestrial intelligence. Since the late 1950s, the prevailing logic has been that aliens might be beaming radio signals into space, so we need only tune to the right frequency to listen in. But our attempts to date have covered just a few thousand stars in a galaxy of hundreds of billions. Perhaps little wonder that we haven't heard a peep.

The latest round started in 2016, courtesy of a ten-year, $100 million project called Breakthrough Listen. Funded by tech entrepreneur Yuri Milner, it will set two of the world's largest radio telescopes surveying the million closest stars across a broader swathe of the radio spectrum, and will cover ten times as much sky as all previous searches combined.

Yet for all its promise this is still mid-twentieth century thinking, as its architects readily admit. Just like steam power on Earth, communicating with electromagnetic radiation could prove no more than a passing craze for a distant civilization. Considering that extraterrestrial societies might be millions of years old, our radio search may be laughably passé.

Now some exciting new alien-hunting strategies are opening up, from scrutinizing the way they breathe to detecting vast feats of cosmic engineering.

The air up there

All life forms we know of are machines for self-replicating, converting fuel into waste as they do so. Over time, the biochemistry of countless billions of individual creatures can transform a world in remarkable ways.

On Earth, virtually all animals inhale oxygen and emit carbon dioxide, while plants spend much of their time doing the opposite. Certain bacteria churn out methane and ammonia. So life here produces a cocktail of gases – the distinctive biosignature of our planet.

When NASA's James Webb Space Telescope launches in 2018, its deep scans of exoplanet atmospheres might give us our first glimpses of biosignatures from other worlds. The telescope will study planets that transit the faces of their host stars. During these minuscule eclipses, background starlight passes through the halo of the planet's atmosphere. It can then be picked apart, trillions of kilometres away, to uncover the spectral fingerprints of molecules.

With luck, hardy biosignatures similar to our own will jump right out. In reality, though, exotic, un-Earthly geological phenomena coupled with novel alien biochemistries could muddy the signal and frustrate our efforts to decode it.

Poison gas

Even if an exoplanet's atmosphere does bear clear signs of life, that won't tell us whether it is inhabited by mindless green slime or sentient city-builders. A more reliable approach to finding kindred spirits is to look for technosignatures – chemicals that only brainy ETs can produce.

Assuming that these intelligent aliens have a basic understanding of chemistry and that – for a while at least – they shared our affinity for generating large-scale pollution, Harvard astronomer Avi Loeb has suggested we target gases such as chlorofluorocarbons (CFCs). Once used in air conditioners and aerosol cans, these nasty, ozone-depleting agents are now being phased out. Yet some can persist in the atmosphere for tens of thousands of years, possibly offering an extended detection window.

For the James Webb Space Telescope to see a CFC signature, levels in an exo-atmosphere would need to be at least 10 times those on Earth. That might be a step too far even for the least environmentally conscious species – unless the denizens of a cold planet decide to load their air with CFCs, using their potent greenhouse effect to trap heat.

City lights

An advanced planetary civilization with night vision as limited as ours might festoon its built environment with artificial lighting. Could we detect alien cities shining on their planet's dark side?

FIGURE 7.3 Japan sends a message to the astronomers of Pluto.

Tokyo could be seen from the distance of Pluto with existing telescopes. Seeing city lights on planets around even our nearest stars would require a space telescope with a mirror 200 metres across. That's around 40 times the size of the one to be fitted in the James Webb Space Telescope, and is unlikely to be on the cards until next century. But maybe Tokyo is small beer next to some alien cities? A planet-wide cityscape might be visible to a space telescope we could hope to build in the near future.

Our attempts might be thwarted by humongous blooms of bioluminescent exo-algae, which could mimic gleaming city-scapes; and like radio communications, cities with artificial light-ing might only be a passing phase, in which case we'd have to get very lucky to catch an alien species at just the right moment.

Aliens on the go

Before too long, humans might be setting up shop elsewhere in the solar system. We might even embark on interstellar missions of exploration. Alien civilizations might already have beaten us to the punch. Setting aside physics-bending hyperdrives and wormholes – whose signals would be hard to predict anyway – what exotic forms of transport should we look for?

If our own cutting-edge technology is any guide, spacefaring aliens might be travelling in modified sailboats, using pressure generated by incoming light. So far, our efforts in this direction have been puny, but scaling up to a football-pitch-sized sail to reap the winds of high-powered lasers is feasible. The leakage of such bright light should be easy to spot with today's telescopes. The Breakthrough Listen project will be complementing its search for alien radio signals with the biggest scan yet for optical lasers being beamed in our direction, indirectly helping us to spot aliens surfing the cosmos.

Less promisingly, aliens might have got as far as building craft powered by nuclear fission and fusion. Their minimal light output would be hard to pick up, unless the star-trekking ETs happened to be right under our noses. A better bet would be aliens capable of exploiting physics as we know it to its limits. They might be using a matter–antimatter annihilating drive, which gives the most bang for your buck in terms of fuel conversion. Such a vessel would spew intense plumes of light. Perhaps tomorrow's gargantuan telescopes might reveal an interstellar rush hour.

Building big

In October 2015, the internet was awash with rumours that an alien civilization had been spotted. The Kepler Space Telescope had seen the light output of a star called KIC 8462 (known informally as Tabby's star) fluctuating by as much as 22 per cent. Normally this would be put down to the eclipsing effect of an exoplanet, but there was a snag. By comparison, Jupiter would block only 1 per cent of our sun's light, and as KIC 8462 is more than half as big again as our sun, whatever was obscuring it had to be colossal. Had we stumbled upon a vast extraterrestrial construction project?

That idea sits nicely with what's known as Dysonian SETI, after the physicist and engineer Freeman Dyson, who laid its foundations in 1960. He suggested we look for evidence of alien macro-engineering: just as the pyramids of Egypt have outlasted their creators, megastructures in space would offer ET-hunters an enduring target.

A classic example is the Dyson sphere, solar-energy collectors encircling a star in a swarm or possibly even a rigid shell. It would harvest a star's energy, ideal for projects such as supercomputers powerful enough to model the entire past and future of the universe. A complete sphere would eclipse its star entirely, but we could still hope to spy waste heat from its construction, or perhaps the funky flickerings of a work in progress, which might look to us like large-scale exoplanet transits.

It turns out that the fluctuations of Tabby's star affect ultraviolet light more than infrared: highly unlikely if large objects are blocking the light. The latest favourite is an uneven ring of dust around the star. But other stars may yet show a clearer signal from alien megastructures.

Feeling the heat

We may think of ourselves as advanced. But a ranking system devised by Soviet astronomer Nikolai Kardashev puts humanity well and truly in its place. In the 1960s, he proposed to rate civilizations based on the energy they were capable of harnessing. One capable of extracting all the available energy from its home planet has K1 status. A civilization milking a tame star with a Dyson sphere would be a K2, and one able to hoover up all the energy in its home galaxy is a godlike K3. On the Kardashev scale, we humans come in at a lowly K0.73.

As impressive as K2 and K3 might be, basic physics suggests their schemes would still cast off prodigious waste heat, probably in the form of infrared light. The most ambitious search for this to date is the Glimpsing Heat from Alien Technologies (G-HAT) survey, which has been hunting for potential K3s using NASA's Wide-field Infrared Survey Explorer (WISE). It was able to whittle down nearly 100 million galaxies to 100,000 infrared-rich eyebrow-raisers. However, further analysis noted that this extra infrared was entirely consistent with vigorous star formation in particularly dusty galaxies. Even so, sub-K3 demigods could be lurking unseen, only reprocessing a small fraction of their galaxy's light.

Alongside star-enveloping solar farms, ET could go in for big science. Alien particle physicists, if any there be, might laugh at our feted Large Hadron Collider, capable of collision energies of 13 teraelectronvolts. The real meat of physics occurs at scales millions of billions times higher than this, energies so elevated that they may at last persuade gravity to unite with the other fundamental forces of nature. A particle accelerator to probe this realm may be unmissable: the size of a galaxy, giving off copious waste heat and radiation.

The most likely particles to be streaming away from these XXLHCs would be neutrinos, tiny bits of nuclear shrapnel that barely interact with normal matter. Yet these neutrinos would be so vanishingly rare in the total cosmic census that today's detectors, like the cubic-kilometre IceCube in Antarctica, wouldn't have a prayer of pinning one down. Instead, we'd need a detection network of billions of sensors arrayed in Earth's oceans. An investment on that scale might be slightly beyond the scope of any one billionaire philanthropist.

8
Mysteries of the Milky Way

Even if your skies are not bleached by city lights, all you can see of our home galaxy is a tousled band of luminescence. It is a faint echo of the galaxy's true majesty: hundreds of billions of stars, some of which make our sun seem like a candle flame, marking out a many-armed spiral; colourful gas clouds and inky dust lanes; and at its heart a few frantic stars and an eerie X-ray glow that betray the presence of a slumbering monster.

Island home

Sitting inside the Milky Way, it is hard to see the wood for the trees. Interstellar dust chokes much of the galaxy and obscures some regions completely. But gradually the giant geography is emerging from the mist, as over the past few decades astronomers have painstakingly mapped our galactic island.

The Milky Way's most prominent feature is a dense disc of stars, gas and dust some 100,000 light years across. A bulge of stars about 12,000 light years across protrudes from the disc's centre, making the whole thing look like two fried eggs back to back. Surrounding these bright parts of the galaxy is a faint spherical halo of old stars, and more than 150 tight-knit balls of stars called **globular clusters**. In total, the Milky Way contains at least 250 billion stars, perhaps as many as a trillion.

This picture began to emerge in 1923 when American astronomer Edwin Hubble became the first person to pinpoint the distances of stars in a distant, fuzzy patch of light called the Andromeda nebula. His measurements revealed that it is in fact an isolated cosmic island of stars. That suggested the Milky Way too is a galactic island.

Spiral sprawl

Observations since then have also shown that our galaxy has bright spiral arms composed of dense concentrations of stars swirling through the disc. We can't travel outside the Milky Way to see what it would look like from afar, but it may well look like a giant Catherine wheel in space. Such large spiral galaxies are unusual. The vast majority of galaxies are small, faint and blob-like, but ours is magnificent.

However, our picture of the galaxy is still patchy. One mystery is how the spiral shape emerges. It doesn't reflect the paths of stars, which follow roughly circular orbits around the galactic centre. Instead, scientists believe that a disturbance called a density wave creates the pattern. Orbiting stars and gas clouds periodically pass into spiral pressure waves that stretch from the galactic centre to the edge of the disc, like cars running into a band of denser traffic. The waves compress gas and trigger the formation of copious bright new stars which highlight the spiral. But it's not clear what triggers the density waves in the first place.

In our own galaxy, the spiral pattern is quite hard to see because it is edge-on and the dusty disc obscures much of our view. Astronomers have done their best to map the positions of the bright stars inside the spiral arms, and radio

FIGURE 8.1 An artist's impression of the Milky Way, viewed from far above the plane of the galaxy.

telescopes have picked up signals from the compressed hydrogen clouds within them. In this way they've traced fragments of four main arms named Scutum-Crux, Sagittarius-Carina, Perseus and Outer. There also seem to be a few smaller arcs between them.

In late 2003, a team led by Naomi McClure-Griffiths at the Australian National University in Canberra announced the discovery of another arm fragment. Using the 64-metre Parkes radio telescope and a separate array of six 22-metre radio dishes, her team spied an arc of dense hydrogen about 77,000 light years long running along the galaxy's outermost edge. Viewed from Earth, the arc appears nearly 150 times as wide as the full moon and is probably part of the galaxy's Outer arm.

Ripple effect

In 2013, a three-dimensional map of the speeds and distances of thousands of stars showed that that the galaxy is undulating up and down. Mary Williams at the Leibniz Institute for Astrophysics, Potsdam in Germany, and her colleagues examined data from the Radial Velocity Experiment (RAVE) survey, which covers almost half a million stars spanning 6500 light years in all directions. The team focused on a category of stars called red clump giants, which have about the same brightness and so are easier to compare when calculating their relative speed and distance from us. They combined the RAVE data on horizontal motion with other readings of how the stars move up and down.

What they found is that stars closer to the centre of the galaxy are spreading outwards above and below the plane, while stars further from the centre are squashing inwards. The motions of individual stars within these zones are chaotic, with some sloshing around in odd directions. But if we could see the overall pattern from the outside, our section of the galactic disc would resemble a flag rippling in the breeze.

It is possible the wave is a lingering effect from a galaxy that smashed into ours in the past. Or it could be actively produced as our satellite galaxies, the **Magellanic Clouds**, spiral around the Milky Way and distort its disc. A more exotic possibility is large lumps of dark matter (see Chapter 9) stirring up the disturbance.

We could soon find out whether the whole Milky Way is waving, with data from the European Space Agency's Gaia spacecraft, which is in the midst of a mission to make a three-dimensional map of the galaxy.

Obscure beginnings

How did the Milky Way form? It's a question that is stubbornly difficult to answer. The galaxy's oldest stars are roughly 13 billion years old, suggesting they formed less than a billion years after the universe began life in a giant explosion 13.8 billion years ago. The big bang created a hot, super-dense fireball that gradually expanded and cooled. But this fireball was not completely uniform; rather, it developed myriad dense patches that somehow seeded the clumpy galaxies we see today.

However, astronomers are vague on the details. Did stars or small star clusters form first, then clump together under gravity to form galaxies? Or did gas and dust in the young universe first form huge structures that only later fragmented to spawn stars?

Desirable neighbourhood

If galactic real-estate agents existed, they'd be keen to market the solar system: a property in a quiet area, with excellent views, rich in chemical commodities, and no trouble from noisy stellar neighbours.

The sun sits in a suburb about 26,000 light years from the galactic centre, just over halfway out. Our suburban locale may be no accident. Some scientists argue that in order for life to evolve on a planet, it probably needs lots of elements heavier than hydrogen and helium to create molecules with diverse biological functions, and these heavy elements are most abundant towards the centre of the galaxy. So life-forming planetary systems can't be too far from the galactic core. On the other hand, advanced life has taken billions of years to evolve, which might be difficult close to the galactic centre among frequent violent supernova explosions.

Clash of the Titans

No galaxy evolves in total isolation. In fact, galaxies are continuously on the move, and the gravitational attraction between them often puts them on a collision course. In a galaxy, stars or star systems are typically a few light years apart. So when galaxies collide, their stars rarely hit each other. Interstellar clouds within galaxies do collide, though, then collapse under gravity, triggering the formation of new stars.

Far from being rare events, there is at least one galactic collision happening here and now. The Sagittarius dwarf **elliptical galaxy** is being gobbled by the Milky Way. On the opposite side of the galactic disc from us, this alien dwarf galaxy contains some 30 million stars, mostly yellowish old ones. It is moving up through the disc at about 250 kilometres per second.

Much larger mergers probably happened in the distant past. Superimposed on the Milky Way's thin disc, which is made of stars of all ages, is a thicker one containing only old stars. A likely explanation is that the Milky Way cannibalized another galaxy about a tenth its size, or several smaller ones, at least 10 billion years ago. The gravitational interaction would have puffed out the stars that had already formed in the Milky Way to make the thicker part of the disc.

One day we will face an even bigger collision. Among the dozens of galaxies that form our **Local Group**, only the Andromeda galaxy matches the Milky Way for size. It is 2.2 million light years away, but every minute the gap closes by about 8000 kilometres. In about 3 billion years, the two giants will run into each other in an encounter that will change them both beyond recognition.

Simulations suggest that over a billion years or so, the galaxies might swing through each other two or three times, stretching out long wispy streams of stars, before settling into a fairly shapeless elliptical galaxy. By this time, the ever-brightening sun will have scorched Earth into a lifeless planet. Maybe humans, or whatever intelligent life succeeds us, will be treated to a spectacular view of Andromeda approaching and filling the sky.

Star thief

Galaxy see, galaxy do. The Milky Way's brightest satellite galaxy stands accused of the same type of crime as the Milky Way itself: tearing apart a celestial object that wandered too close.

The Large Magellanic Cloud is the most luminous of more than 50 galaxies orbiting our own. Lying 160,000 light years from Earth, it is so bright you can see it with the naked eye.

In 2016, Nicolas Martin of the University of Strasbourg in France and his colleagues spotted what looks like a globular cluster – a tightly packed group of stars – in distress. The cluster is on the outskirts of the Large Magellanic Cloud, about 42,000 light years from the galaxy's centre.

It has become stretched out, with its long axis pointing right at the Large Magellanic Cloud. This suggests the galaxy's gravity is pulling harder on the cluster's near side than its far side, yanking it apart.

Still, if the star cluster has been orbiting the galaxy for a long time, the team say it's strange that the destruction is occurring only now. So they suspect the cluster originally circled a less massive galaxy nearby – the Small Magellanic Cloud – with weaker gravity that didn't tear it apart. Only recently did the Large Magellanic Cloud snatch the cluster and begin shredding it.

Sulphurous cloud

You would need a super-sensitive, gargantuan tongue to taste it, but Sagittarius B2, a **molecular cloud** about 100 light years across near the centre of our galaxy, would taste terrible.

Blended into three million suns' worth of hydrogen and helium gas are hints of sweetness: ethylene glycol, the syrupy and toxic mainstay in antifreeze, and ethyl formate, which has a fruity, lemony scent. It's got acetic acid, too, for a bit of vinegar. And there's plenty of booze. The first ethanol found in space was detected there, way back in 1975.

All of this is accompanied by less palatable stuff, like acetone, which works well as nail polish remover, and hydrogen sulphide, the unmistakable stink of rotten eggs. These chemicals are all vanishingly rare compared with hydrogen molecules. But because the cloud is so huge, with thick patches illuminated by brilliant young stars, even trace amounts of different substances leave detectable spectral signatures.

For more than 40 years, that's made Sagittarius B2 a sort of Valley of the Kings or Burgess Shale for cosmic chemists: a place to return to again and again to make discoveries. In doing so, we might gain some insight into the origin of life.

An interstellar organic chemistry lab needs only a few ingredients. In clouds like Sagittarius B2 – and also in the cloud that long ago gave birth to our solar system – icy coatings can condense around tiny grains of dust.

When radiation hits the ice-covered grains, it produces free radicals, driving chemical reactions that can build larger molecules. Stars also condense out of the cloud, as is happening right now in the north fragment of Sagittarius B2. Their light warms the ice, vaporizing whatever molecules have been formed. Out in space, their chemical bonds spin and flex enough to emit radio waves we can detect.

That part of the cloud has offered up an intriguingly intricate series of organic molecules. In 2008, a team led by Arnaud Belloche at the Max Planck Institute in Bonn, found amino acetonitrile, a close relative of glycine, the simplest amino acid. In 2014, the same group announced the first detection of an interstellar molecule with a branching carbon backbone. That suggests complex amino acids might also be able to grow in space. And in 2016, another team found the first chiral interstellar molecule – a structure, common in biology, that can come in different mirror-image versions like your right and left hands.

In our solar system, comets seem to have many of the same molecules, even amino acids. So do meteorites. It's possible that these substances, delivered through crash landings on early Earth, provided some of the ingredients life needed to start. But before that – long before – they may have been grown in the thin, icy rind of irradiated dust grains, drifting in space, bathed in the light of newborn stars.

Dark heart

At the heart of the Milky Way lies a monster. By watching the speeds of stars flying around in the galactic centre, astronomers have shown that it harbours an invisible, dense mass, about 4 million times as massive as the sun. It is concentrated in such a small area that the only plausible explanation seems to be a single **supermassive black hole**.

Most, if not all large galaxies harbour a central black hole. Some of these behemoths are enthusiastic eaters, pulling in surrounding gas with abandon. As the gas falls towards the black holes, it heats up, producing blazing beacons known as quasars that can be seen across the universe (see Chapter 9). But our own supermassive black hole, known as Sagittarius A*, looks unusually sleepy and dim, producing only a modest glow of X-rays and radio waves.

That is despite apparently having plenty of material to gobble. A crowded disc of massive stars spins around it, and researchers had previously calculated that these stars should spew out enough gas in stellar winds to provide the black hole with about four Earths' worth of meals over the course of a year. If it were swallowing that much material it would shine 100 million times brighter in X-rays.

That may be because the gas around Sagittarius A* is so hot. Collisions between stellar winds in the starry disc heat the gas to 10 million °C before it even starts to fall towards the black hole. This hot gas is tenuous and its particles zip around randomly, making it hard to corral. Jets of matter emerging from near the hole may also play a role, blowing away gas that might otherwise fall in.

Old flares

The feebleness of our supermassive black hole may be just a passing phase. In 2003 a satellite called Integral revealed extremely energetic X-rays coming from the gas cloud Sagittarius B2, about 350 light years from the black hole.

A plausible explanation is that to observers on Earth around 350 years ago, the galactic centre's black hole would have looked incredibly bright at short wavelengths – a million times brighter than it looks today. If Newton or Galileo had invented a first-rate gamma-ray telescope, they might have seen a dazzling display from the galactic centre. The bright radiation reached the B2 cloud 350 years after the flare-up, and we now see the cloud fluorescing in X-rays as a result. This relatively recent stint as an active galaxy suggests it might well flare up again in future, but there is no telling when.

A far bigger flare-up seems to have happened 2 million years ago. In 2010, astronomers using NASA's Fermi gamma-ray satellite spotted a pair of spectacular but mysterious structures now called the Fermi bubbles, towering 25,000 light years above and below the galactic plane. Theories to explain these enigmatic objects range from gamma rays emitted by annihilating dark matter to supersonic winds unleashed by intense bursts of star formation.

Then in 2013 Bill Mathews of the University of California, Santa Cruz, and Fulai Guo of the Shanghai Astronomical Observatory in Beijing, argued that the bubbles were caused by an outburst from Sagittarius A*. The idea is that as the supermassive black hole pulls matter in, this matter accretes in a surrounding disc, heats up and starts glowing. When large amounts of matter get pulled into the disc, energy is released as bright jets of particles perpendicular to the black hole's spin. Simulations showed that two such jets of high-energy particles could have created the bubbles. The flare-up, they calculated, would have happened between 1 and 3 million years ago and lasted a few hundred thousand years.

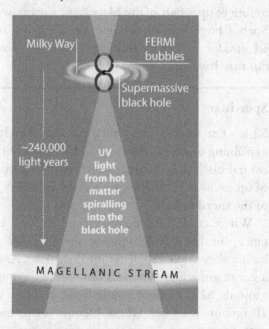

FIGURE 8.2 Two million years ago, our galaxy's black hole may have created two bubbles of gas and lit up part of an interstellar gas cloud.

Joss Bland-Hawthorn of the University of Sydney realized that such an outburst might solve another longstanding mystery. In 1996, astronomers discovered that a section of the Magellanic stream – a fast-moving flow of mainly hydrogen gas about 240,000 light years from the Milky Way – is glowing about 10 to 50 times as brightly as the rest. Could the same explosion that blew up the Fermi bubbles be responsible? After all, the bright part of the stream lies below the galactic centre.

Based on data from other galaxies with supermassive black holes that are actively spewing jets, Bland-Hawthorn and his colleagues worked out that if Sagittarius A⋆ had been similarly active, the resulting UV light would indeed have ionized – and therefore lit up – part of the Magellanic stream (see Figure 8.2).

Such a huge flare-up would have formed a bright moon-sized smudge in Earth's southern sky. *Homo habilis* or *Homo erectus* may have had a ringside view.

Speedstar

S2 is a fast, intense, blue-white star that frankly has some explaining to do. It orbits within a whisker of the galaxy's central black hole, Sagittarius A⋆, swinging by at a speed of up to 5000 kilometres per second, or nearly 2 per cent of the speed of light.

Where could it have come from? At such a short distance, the black hole's gravity should shred gas clouds before they could condense into new stars. And although a star might migrate inwards from more tranquil breeding grounds, S2 is a bright young thing no more than about 10 million years old, whose lifetime seems too brief for such a trek.

Interview: How I'm going to photograph a black hole

Radio astronomer and astroparticle physicist Heino Falcke is based at Radboud University in Nijmegen, the Netherlands. He plans to use a global network of radio telescopes to snap the black hole at the Milky Way's heart. New Scientist *interviewed him in 2015.*

Why photograph a black hole?

Black holes were predicted a century ago, but I have the feeling that we understand them even less these days. We still don't have conclusive evidence for the presence of an event horizon – their point-of-no-return surface. Also, event horizons and quantum theory just don't go together. Something needs to change, and it's not entirely clear what that is.

How do we even know there's a black hole in the Milky Way's core?

Stars in the galactic centre orbit at some 10,000 kilometres per second, meaning there must be a central mass that is more than 4 million times our sun's mass. The only thing that we see in the very centre is a source of very short, sub-millimetre radio waves called Sagittarius A★.

How will your planned giant network of radio telescopes help?

The black hole's event horizon is probably 25 million kilometres across, but it's 27,000 light-years away. To image it at sub-millimetre wavelengths you need a telescope as big as Earth. A worldwide network of radio telescopes can obtain the same resolution.

Aren't US astronomers working on a similar idea?

I first discussed these ideas ten years ago with Shep Doeleman of the Massachusetts Institute of Technology, who now heads the US-led Event Horizon Telescope project. It makes no sense for us or them to work with a subset of the available telescopes. We need each other.

What exactly are you looking for?

We hope to see how radio waves from the black hole's surroundings are bent and absorbed, just as in Christopher Nolan's movie *Interstellar*. The result should be a sort of central 'shadow'. By comparing the size, shape and sharpness of this shadow with theoretical predictions, we can test general relativity. If the shadow is half as big – or twice as large as predicted, say, general relativity can't be correct.

What are the biggest challenges?

The technology is daunting, but now under control. For each telescope you have to record hours of data at a rate of 64 gigabits per second and ship hard discs with petabytes of data between continents. Budget issues have eased a little bit with grants from the European Research Council and the US National Science Foundation.

When will we have our first black hole portrait?

In 2000, I said a result might be in within a decade, so I'd better temper expectations a little bit. Will it be another tem years? I hope not, but in the end it takes the time it takes.

Boson star

Fascinating, bamboozling, vaguely terrifying: black holes are the love-to-hate monsters of the universe. These insatiable cosmic cannibals are concrete predictions of Einstein's general theory of relativity, the best theory of gravity we have. Soon we should have the first direct image of a black hole – the big one at the Milky Way's centre. But what if it isn't there?

Our obsession with black holes might have blinded us to the existence of something even stranger – a basic phenomenon of particle physics whose significance we have failed to grasp. It is a speculative idea as yet, to be sure, but there are sound reasons to contemplate it.

No one knows how black holes work on the inside. They represent the point where the very large, the domain of general relativity, meets the very small, the domain of quantum theory – and the results are not pretty. Relativity suggests that anything that falls in will be crushed by the black hole's gravity into a singularity of zero volume and infinite density, but that makes a nonsense of any equation. Meanwhile, theorists' most refined calculations show that black holes must either destroy information – a complete no-no in quantum theory – or surround themselves in a seething mass of energy called a firewall, which breaks a tenet of general relativity.

Strong evidence for black holes came in 2016 with the announcement of the first detection of gravitational waves (see Chapter 10). The signals observed by the LIGO and Virgo experiments were exactly those predicted for the collision and merger of two stellar-mass black holes.

Case closed? Not so fast, says Luciano Rezzolla of the Frankfurt Institute of Advanced Studies in Germany. Those signals might not come from black holes, but from an entirely different theoretical invention: boson stars.

The fundamental particles that make up most matter all belong to a class known as fermions. Their signature characteristic is that they obey the Pauli exclusion principle, which says that particles cannot occupy the same quantum energy state as one another. The Pauli principle explains the appearance of the material world: it determines how electrons arrange themselves in different energy states around an atomic nucleus, and thus the properties of the various chemical elements.

Bosons are a different kettle of fish. The Higgs boson, discovered to great fanfare in 2012, is perhaps the most famous example. It provides matter particles with their mass; other bosons carry the forces that allow matter particles to interact. Bosons aren't exotic. In fact, we see them all the time, quite literally: photons of light are bosons that carry the electromagnetic force.

The thing about bosons is that they can cram together with virtually no limits. They become what is in effect a collective particle, a state of matter known as a Bose–Einstein condensate. We can make Bose–Einstein condensates in the lab. We now know that, given the right bosons, there's nothing to stop them forming something on a bigger scale – perhaps much bigger. Some physicists even think Bose–Einstein condensates can form stars, although not as we know them.

When normal matter forms a star, gravitational pressure heats it so it ignites into nuclear fusion, pouring out light. In contrast, boson stars would just hang there like cosmic couch potatoes. Doughnut-shaped couch potatoes: simulations suggest that if boson stars rotate as conventional stars do, centrifugal forces would give the bosonic matter that form.

These celestial doughnuts would be transparent. Emitting no light of their own, they would be invisible, and the primary thing that would give them away would be their intense gravity. Sound familiar?

The idea of boson stars isn't new, but astrophysicists pooh-poohed it because no one could think what sort of boson might be used to make one – the particles such as photons that transmit the fundamental forces don't work. Then came the discovery of the Higgs, which revived interest in novel bosons. A prime candidate for forming a boson star is the axion, an ultralight hypothetical particle that could also form dark matter, the mysterious glue that astronomers believe holds galaxies together.

How can we seek evidence of boson stars? Gravitational waves could help. In the aftermath of a merger, when the two objects have coalesced and are still quivering from the shock, a new boson star will have a different frequency from a black hole. After an upgrade in five years or so LIGO may be able to hear this quiver with enough precision to tell the difference.

The Event Horizon Telescope might deliver clarity sooner, although opinions differ as to whether it will be easy to distinguish between images of a black hole and a boson star. Calculations by Frédéric Vincent of the Paris Observatory in France suggest that the gravity of a compact boson star will bend light around itself, creating an empty region that could be mistaken for the shadow of a black hole event horizon.

Rezzolla thinks this analysis is overly pessimistic. Like a black hole, a boson star will be sucking in matter from its surroundings, but the boson star's transparency means this matter will be visible at its centre. It is also likely to heat up and start emitting light or some other form of electromagnetic radiation.

The reward for killing off black holes is potentially immense. Embodying the conflict between general relativity and quantum theory as they do, they are a massive roadblock to progress on an overarching theory of nature.

So what is dark matter?

We know that it can't simply be ordinary matter, based on protons, neutrons and electrons, somehow hidden away by a dark coating. If the universe held very much more ordinary matter than we can account for today, then when the early universe was a few minutes old, densely-packed protons and neutrons would have fused to form a higher proportion of helium than we see. Dark matter has to be something else.

The usual guess is that it is some kind of exotic particle that only interacts weakly, so light goes straight through it. The established standard model of particle physics doesn't hold anything that would fit the bill, but various proposed extensions to the theory do. The options include the long-time leader, a broad class of hefty particles called WIMPs (for 'weakly interacting massive particles') and much lighter axions.

Dark matter might not be particles at all – instead it could be in the form of primordial black holes, created in the early moments of the universe.

9
Galaxy quest

The vast Milky Way is only one among many billions of galaxies, from insignificant smudges of just a few million stars up to giant ellipticals. We see some in grand, slow-motion collisions; others in the dense swarms of galaxy clusters, giving hints at the nature of dark matter. And almost all of them harbour a supermassive black hole.

Local matters

Our local group of galaxies is a bucolic spot, barely a village in cosmic terms. Within a few million light years there are three spiral galaxies – the Milky Way, Andromeda, and the small spiral of Triangulum – plus scores of dwarf galaxies.

According to recent research, the galaxies of the local group may have exchanged much of their material through galactic winds, powered by supernova explosions. Recent models show that for large galaxies, these winds could ferry in about 50 per cent of the matter present today, from distances up to a million light years away. Large galaxies like ours tend to snatch away matter from their smaller neighbours like the nearby Magellanic Clouds. Half of the atoms in your body could be intergalactic interlopers.

To leave our village and reach the nearest town, you must travel 60 million light years. The Virgo Cluster is an aggregation of more than a thousand galaxies. As in other large clusters, the galaxies in Virgo bathe in a thin, superhot gas, at a temperature of about 30 million kelvin. Among them is a giant elliptical galaxy, M87, perhaps 100 times the mass of the Milky Way. Unlike spiral galaxies, which have a healthy supply of cold gas that can condense to form bright new stars, ellipticals are relatively dead, with few new stars forming.

Clusters such as Virgo are arguably the largest things in the universe, depending on your definition of 'thing'. We can plot out much larger structures called superclusters and voids, these are simply patterns on the sky, slowly being stretched out as the universe expands. Clusters by contrast have their own independent identity. They are gravitationally bound, meaning that galaxies buzz around on orbits within the cluster, and the whole thing is held against universal expansion.

The rapid motions of galaxies in the Coma cluster, about 300 million light years from Earth, puzzled Fritz Zwicky when he examined them in 1933. They are too fast to be held there by the gravity of visible matter, prompting Zwicky to suggest that some sort of dark matter was present – an idea bolstered decades later by Vera Rubin's observation that, by the same reasoning, the outer reaches of spiral galaxies rotate too fast. It's now thought that dark matter helped galaxies to form, providing the gravity needed to draw in gas. But some baffling galaxies appear to be dominated by dark matter – the record holder, called Dragonfly 44, is about the same mass as the Milky Way, but has perhaps a hundredth of our stock of stars, consisting of 99.99 per cent dark matter. Astronomers do not yet know how these dark galaxies arise.

Kamikaze galaxy

Don't mess with the Milky Way. A dim galaxy in the constellation Hercules has learned this lesson the hard way after diving into our galaxy and being torn asunder by its gravitational pull.

The Hercules dwarf is now 460,000 light years from Earth, nearly three times farther than the Milky Way's brightest satellite galaxy, the Large Magellanic Cloud. Hercules's stars are spread over a great expanse of space, which suggests the Milky Way's gravitational pull has yanked them away from one another.

Andreas Küpper and Kathryn Johnston at Columbia University in New York and their colleagues have used the observed positions and velocities of the galaxy's stars to deduce Hercules's plunging path through space. They calculate that at its farthest the dim galaxy voyages 600,000 light years from the Milky Way's centre. At its closest it's just 16,000 light years out.

That's closer than any other satellite galaxy is known to come. It's even closer than we are.

And according to computer simulations, the galaxy's last passage proved fatal. As Hercules neared the Milky Way half a billion years ago material from our galaxy invaded the dwarf. The gravitational force of this additional matter pulled the galaxy's stars and dark matter towards its centre. Then, as Hercules skirted away from the Milky Way's centre, the stars and dark matter rebounded, causing it to explode into its present distended state. Today, Hercules's stars owe their allegiance to the Milky Way rather than to one another, because they are so dispersed they no longer feel much of their siblings' gravitational pull.

Still, the stars continue moving on similar paths, like parachutists jumping from the same aeroplane. Johnston reckons the next time the galaxy ventures towards us, you might be able to find it as a stream of stars zipping through at very high speed.

Metamorphosis

In 2014 the Hubble Space Telescope spotted half a dozen spiral galaxies that are being ripped apart and remade into jellyfish – with blobby bodies and glowing tendrils of stars – as they move towards joining galaxy clusters. The discovery more than doubled the number of known jellyfish galaxies, and should help researchers better understand how galaxies can transform.

We know that clusters contain many more elliptical galaxies than spirals, hinting that newcomer spirals are somehow being transformed. Jellyfish galaxies may capture the process in action, according to Harald Ebeling at the University of Hawaii in Honolulu.

FIGURE 9.1 A spiral galaxy that strays too near to a cluster can end up as a celestial jellyfish like this.

The space between the galaxies in a cluster is laced with gas, which reaches searing hot temperatures. That means when new galaxies join a cluster, they can't just slip in quietly. Hot gas in the cluster smacks into cold gas within a new arrival, blasting the cold gas outwards in long streams. The stripped body of the newcomer settles into a blobby shape, while cold gas in the tendrils compresses enough to ignite new stars.

Ebeling and colleagues unexpectedly caught their first jellyfish in late 2005 and have been hunting for more extreme examples in Hubble images since then. Until 2014 though, the transformation had been spotted only a few times in relatively nearby clusters. That's probably because the change is over too fast. And once a spiral galaxy has been stripped of its cold gas, it won't experience another drastic transformation.

Studying jellyfish galaxies in detail could help solve another mysterious feature of galaxy clusters: why they contain relatively young orphan stars that do not belong to any particular galaxy. The gas inside clusters is too hot to collapse into new stars, so the stars must come from outside – possibly from the tentacles of jellyfish.

Spiders in the web

Like cosmic spiders, dwarf galaxies have been caught feasting on blobs of gas spread across a hidden web. The very same process seems to have fuelled the birth of stars and the growth of galaxies when the universe was young.

The idea that galaxies grew fat by eating from the cosmic web – a giant mesh beaded with clouds of cold gas – has been around for a while, and recent simulations have increased its popularity. As the gas clouds fall into the gravitational clutches of a galaxy, they spark bursts of star formation, the models show. But the process has been hard to observe. The Milky Way and most galaxies that we can see nearby are filled with hot gas that warms up approaching material, preventing it from collapsing into stars. And because clumps of cold gas in intergalactic space don't emit much light on their own, they're hard to spot.

In 2015, a team led by Jorge Sanchez Almeida of the Astrophysics Institute of the Canary Islands, looked at a set of small, faint galaxies with a low proportion of elements heavier than hydrogen and helium. They were able to infer how oxygen levels varied across these galaxies' discs. They found that the bright, star-forming regions had only about a tenth as much oxygen as was found elsewhere in these galaxies. This was a sign of newly arrived gas powering star formation, they concluded: it had to be recently added, because any old gas would lose its distinctive chemical signature within a few hundred million years through being stirred up into a homogenous gloop. Blobs of gas moving along the cosmic web could explain the new activity, and the very existence of these galaxies.

Cosmic colliders

Galaxy clusters gave us the first hint of dark matter, and today we are poring over collisions between clusters to try and work out what this mysterious material actually is.

Dark matter is thought to make up around 83 per cent of the matter in the universe, but apparently it refuses to interact with ordinary matter except through gravity. Some researchers have tried to do away with dark matter by modifying the laws of gravity instead, But observations of the Bullet cluster, a collision between two galaxy clusters about 3.7 billion light years away, suggest that won't work. Normally, dark matter and ordinary matter are too well mixed to tell them apart. But when the two clusters collided, their galaxies glided past each other and left a trail of hot, interacting gas behind. The dark matter, seen indirectly by its gravitational effects, remained with the galaxies. This suggests that dark matter is a substance rather than an adjustment to gravity, and that its particles don't bounce off each other like atoms and molecules do.

Studies of other galactic smash-ups have hinted that dark matter might interact with itself via a new, dark-matter-only force. Then in 2015, new observations of 30 colliding galaxy clusters with NASA's Chandra X-ray Observatory showed that when galaxies collide, dark matter carries on its path unimpeded and unaffected by any other dark matter around, suggesting that it does not interact with itself after all. But the result does not rule out a very subtle new force. Astronomers will look closer at the nature of dark matter by analysing thousands of other colliding clusters.

Bright black

A vast cloud of cold gas was floating in the void of space, a patch of inert blackness against the even deeper blackness behind. Then, as if from nowhere, a thin jet of matter streaked towards it. The jet slammed into the cloud, compressing its matter and triggering a firestorm of star formation. What had once been a dormant gas cloud was now a galaxy. Could this be how a galaxy is born – sparked by the ejecta from a super-massive black hole?

The link between black holes and galaxies – and specifi-cally which comes first – has puzzled researchers since the mid-1960s, when quasars were discovered. These objects can release up to 100 times as much light and other radiation as the whole Milky Way galaxy, from a region no bigger than our solar sys-tem. Such prodigious output from so small an object can be explained if a black hole millions to billions of times the mass of our sun is pulling in surrounding gas and dust. This material swirls into a disc and heats up to release intense radiation to outshine everything else. There are several other types of **active galactic nucleus** such as radio galaxies and Seyfert galaxies, less luminous than quasars but still pretty powerful, and all of them seem to be the result of feeding black holes.

For some time, many researchers thought that supermassive black holes were found only in such rare active galaxies, mean-ing that they were interesting but inconsequential for the cos-mos as a whole. That view changed, however, with the discovery that most, if not all, galaxies harbour a supermassive black hole. They had remained hidden because in most galaxies, including our own, the hole is starved of fuel and therefore lies dormant.

And there is a clear relationship between these objects and their parent galaxies. Spiral galaxies have central bulges of stars, which are their most ancient large component – and these bulges are consistently about 1000 times as massive as the black holes at their hearts. The same goes for elliptical galaxies, which are all bulge. This suggests that the growth of one influenced the other. Further evidence of a link comes from the fact that star formation and quasar activity both peaked around the same time, 8 to 10 billion years ago.

Even supermassive black holes are lightweight compared with galaxies around them, so you might not expect them to hold much sway. Perhaps the bulge forms first, and then the black hole grows to a size determined by the size of the bulge? But black holes can punch above their weight. Some theorists suggest that bulges and black holes grow at the same time, and eventually when the black hole gets large enough, winds and radiation from its disc blow away any remaining gas, setting a limit on the size of the bulge.

There's a more radical possibility. The black holes in many active galaxies squirt out matter in two oppositely directed, thread-like jets (see Chapter 8). These jets can travel close to the speed of light, and break out of the surrounding galaxy to travel millions of light years into intergalactic space.

In 2009 a team led by David Elbaz at the French Atomic Energy Commission in Saclay reported results on an unusual quasar called HE0450-2958 and its jets. Five billion light years away, HE0450-2958 was the only known naked quasar – a supermassive black hole without a surrounding galaxy.

Using the infrared instruments on the Very Large Telescope in Chile, the team made a startling discovery. The quasar's jets stabbed like a laser beam into a galaxy 23,000 light years away. That galaxy is rich in bright, young stars and is forming them at a rate equivalent to 350 suns per year, a hundred times more than you would expect for galaxies in that area.

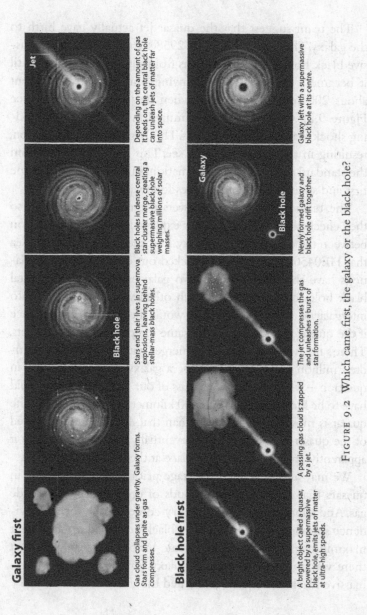

Galaxy first

Gas cloud collapses under gravity. Galaxy forms. Stars form and ignite as gas compresses.

Stars end their lives in supernova explosions, leaving behind stellar-mass black holes.

Black holes in dense central star cluster merge, creating a supermassive black hole weighing millions of solar masses.

Depending on the amount of gas it feeds on, the central black hole can unleash jets of matter far into space.

Black hole first

A bright object called a quasar, powered by a supermassive black hole, emits jets of matter at ultra-high speeds.

A passing gas cloud is zapped by a jet.

The jet compresses the gas and unleashes a burst of star formation.

Newly formed galaxy and black hole drift together.

Galaxy left with a supermassive black hole at its centre.

FIGURE 9.2 Which came first, the galaxy or the black hole?

The team suggest that the quasar jet actually gave birth to the galaxy. They think HE0450-2958 started off as a supermassive black hole that sucked in gas from intergalactic space until it became a quasar. It kept growing until a critical moment about 200 million years ago when its jets switched on (see Figure 9.2). One of the jets slammed into a gas cloud, sending shock waves through the gas. This triggered star formation, resulting in the galaxy we now see. This is a drastic shift from the standard view of galaxy formation, in which galaxies came first and supermassive black holes followed.

The team knew how controversial their idea would be, so they checked there was no other explanation for the association between the quasar and the galaxy. First they considered the idea that HE0450-2958 had been kicked out of the galaxy. Simulations have shown that when two galaxies merge, their central black holes can ricochet off each other, with one ejected into intergalactic space. However, it would be an odd coincidence if the quasar was ejected in the same direction as one of its jets. There's no evidence of a galaxy merger within the past few hundred million years. And to escape a galaxy as big as the one in question – which is about the mass of our Milky Way – it would have to be kicked out at about 500 kilometres per second, but the quasar is moving much slower than that. In fact, the low speed of the quasar means that it will eventually fall into the galaxy it apparently created, and take its place at the heart of the galaxy.

We may also have seen the stage prior to star formation, in quasars that are offset from clouds of cold carbon monoxide gas. And support for Elbaz's theory comes in the form of evidence showing that supermassive black holes had reached their maximum mass early on in cosmic history. This suggests that there was an epoch preceding galaxy formation when supermassive black holes grew and ruled the universe.

If so, where did they come from? Observations of quasars show that monster black holes weighing 10 billion solar masses formed within a billion years of the big bang. For years, astrophysicists have been puzzling over how they could have grown into such behemoths so quickly.

One idea is that they grew from the much smaller black holes which form when a star reaches the end of its life and collapses. In a superdense cluster of stars, several of these black holes might merge to create a huge one that continues to grow by feeding on gas. But critics of this idea point out that there was simply not enough time in the first billion years after the big bang for stellar-mass black holes to merge into something big enough.

An alternative involves the formation of supermassive stars. If a star like this ever formed, it would be so massive that the heat generated by nuclear burning at its core would not be enough to oppose the gravity trying to crush it. The whole thing would collapse at once, creating a supermassive black hole. Mitchell Begelman at the University of Colorado in Boulder has studied this picture in detail. According to his calculations, the seeds of supermassive black holes formed inside the supermassive stars, and eventually the outer layers of these curious stars exploded to reveal the black hole hidden within. To test the idea, we'll have to wait for the next generation of telescopes.

Attack of the green blobs

Imagine making a night-time trek to a remote stretch of desert, far from any sign of civilization. You crest a hill and are astonished to find a building ablaze with artificial lighting.

That is a little like the puzzlement that greeted the discovery of Hanny's Voorwerp, a curious gas cloud found floating in intergalactic space in 2007. It is brighter than 30,000 suns but has no obvious power source. Now, 19 similar clouds have been discovered, all glowing apparently without internal power.

The clouds were probably energized by nearby monster black holes that had blasted them with intense radiation. This link to huge black holes is exciting because it means the clouds could be an excellent new way to probe the growth and feeding habits of these inscrutable behemoths.

Hanny's Voorwerp was discovered by the schoolteacher Hanny van Arkel while she was classifying galaxies as a volunteer for the Galaxy Zoo citizen science project. She noticed a weird blob that appeared intensely blue in the false-colour image she was examining and emailed the Galaxy Zoo researchers about it. Voorwerp means 'thing' in Dutch, Van Arkel's native language.

Intrigued, the researchers scheduled new telescopic observations of the object. The object's light spectrum shows that its glow comes from oxygen that has been ionized – stripped of some of its electrons – making its true hue a greenish colour. Other elements were ionized too. It would take a huge amount of energy to ionize all this gas, but there was no hint of a source. Radiation from hot young stars could account for ionized oxygen in the cloud, but not the ionized neon: neon doesn't shine in the ultraviolet, as seen in the cloud, without lots of X-rays hitting it.

That suggested a monster black hole was involved. Most galaxies are thought to host one in their core and in many cases matter spiralling into the black holes produces huge amounts of X-rays.

A galaxy called IC 2497 lies about 45,000 to 70,000 light years from the glowing cloud, and a black hole at its core could easily blast Hanny's Voorwerp with X-rays. But there's a catch. IC 2497's core shows no sign of emitting X-rays.

In 2008, the Galaxy Zoo team concluded that less than 100,000 years before IC 2497 became the galaxy we see today, its black hole was gulping down a big meal and sending out a torrent of X-rays. Because it takes time for the X-rays to reach the cloud, some of them were still arriving and making it glow when it emitted the light Van Arkel saw, even though the black hole was by then quiet.

That's a rare bit of evidence of how much black-hole feeding can vary over tens of thousands of years. Researchers are keen to understand the feeding habits of black holes because such binges, called accretion events, have an enormous effect on their surroundings, shutting off galaxy growth by heating and expelling the gas needed to form new stars.

But it was not clear how representative Hanny's Voorwerp was of black-hole behaviour. Since then, professional researchers and Galaxy Zoo volunteers working together have found many similar objects – glowing gas clouds near galaxies whose black holes appear quiet but probably blasted the clouds in the past.

Most of the newly discovered clouds have a nearby galaxy that is interacting or merging with another galaxy. That fits with the black-hole-blast explanation, because such encounters tend to shake loose gas clouds that then stray into intergalactic space – providing targets to be illuminated by black-hole X-rays.

It also shows Hanny's Voorwerp is not a freak. All over the universe, black holes are apparently firing broadsides at their surroundings, then quickly quietening down and fading from view, like flashing lights.

Cosmic web

As astronomers survey the skies with ever more sensitive telescopes, they have begun to discern the texture of the universe. Today's galaxies and galaxy clusters, vast in themselves, are arranged into big strings and knots called superclusters. Sometimes these in turn can be linked together into features called walls. On the grandest scales, the universe resembles a cosmic web of matter surrounding relatively empty voids. It has been likened to a foam.

This webby/foamy structure matches computer simulations, which take their starting point as an almost smooth universe, where dark matter outweighs the gas of ordinary matter (made of neutrons, protons and electrons) about five to one. Mapping voids could give us a new way to probe the repulsive stuff known as dark energy (see Chapter 10), which will affect their apparent shape. So far so good.

But as our observations of the cosmos come into sharper focus, astronomers are beginning to identify structures bigger than any seen before. In the nearby universe, we know of the Sloan Great Wall, and in 2014 the Milky Way was found to be part of a supercluster system called Laniakea. Both of these structures are enormous. Then in 2016 astronomers traced out the BOSS Great Wall, around 5 billion light years away and with a total mass perhaps 10,000 times as great as the Milky Way. It is two-thirds bigger again than either the Sloan wall or Laniakea. It contains 830 galaxies we can see and probably many more that are too far away and faint to be observed by survey telescopes.

What's the biggest galaxy?

According to the standard model of galaxy formation, the biggest galaxies are elliptical monsters formed from the collision of many smaller galaxies. The largest known example is the lens-shaped IC 1101, a billion light years away in the centre of the Abell 2029 galaxy cluster. IC 1101 is close to 6 million light years across, making it thousands of times the volume of the Milky Way.

After Copernicus

These growing walls, voids and other hints of giant things are proving troublesome. Ever since Copernicus proposed his revolutionary idea that Earth's place among the stars is nothing special, astronomers have regarded it as fundamental. It evolved into the cosmological principle: nowhere in the universe is special. There are patches of individuality on the level of solar systems, galaxies and galaxy clusters, of course, but zoom out far enough and the universe should be homogeneous. No vast hyperclusters or voids bigger than about a billion light years. Conveniently this assumption of smoothness makes it much easier to use Einstein's general theory of relativity to model the universe as a whole.

It doesn't sit well with the giant hole in the universe – a void almost 2 billion light years wide – identified in 2015 by István Szapudi at the University of Hawaii at Manoa and his colleagues. The team call this vast expanse a supervoid, and believe it might explain away the giant cold spot seen in maps of the cosmic microwave background, an observation that has been puzzling astronomers for more than a decade.

And the supervoid isn't the half of it. As far back as 2012, a team led by Roger Clowes at the University of Central Lancashire, UK, claimed to have found an enormous structure strung out over 4 billion light years – more than twice the size of the supervoid. This time it wasn't an empty patch of space, but a particularly crowded one. Known as the Huge Large Quasar Group, it contains 73 quasars – the bright, active central regions of very distant galaxies. Astronomers have known since the early 1980s that quasars tend to huddle together, but never before had a grouping been found on such a large scale.

Then in 2015, a team of Hungarian astronomers uncovered a colossal group of gamma-ray bursts (see Chapter 10). The galaxies emitting these bursts appear to form a ring 5.6 billion light years across – 6 per cent of the size of the entire visible universe.

Intruders from another dimension

The solidity of both this burst ring and the quasar group have been disputed, but some researchers believe that these cosmic megastructures point to something fundamentally new. One is Rainer Dick, a theoretical physicist at the University of Saskatchewan, Canada. He has made a breathtakingly audacious proposal: these megastructures are the first evidence of other dimensions intruding into our own, leaving dirty footprints behind on our otherwise smooth and homogeneous cosmic background.

For decades, many theorists have regarded the existence of extra dimensions as our best hope of reconciling Einstein's general relativity with that other bastion of twentieth-century physics: quantum theory. A marriage between these two seemingly disparate concepts – one dealing with the very large and the other with the very small – would yield what is often called a theory of everything, a one-size-fits-all framework capable of describing the universe in its entirety. One popular candidate is M-theory, an extension of string theory that famously suggests we live in an 11-dimensional universe, with the seven dimensions we don't sense curled up so tightly as to drop out of sight. It's an elegant and mathematically appealing framework with a number of influential supporters – but it lacks solid predictions offering opportunities to verify it. Dick's work on a generalization of string theory known as brane theory might provide just such a prediction.

At the heart of brane theory is the idea that what we perceive as our universe is a single four-dimensional membrane floating in a sea of similar branes spanning multiple extra dimensions. And according to Dick's calculation the effects of a neighbouring brane overlapping with ours could skew measurements of distance using **redshift**, producing vast mirages that could explain supervoids and other megastructures.

Redshift, blueshift

One measure of distance to far-off objects is through an effect known as cosmological redshift. The light from a distant object has been travelling for a long time through expanding space, which stretches it out to longer, redder wavelengths. Astronomers break down the light from the object using a spectrometer to reveal distinctive spectral lines. The further away the object, the faster it will appear to recede and the more the lines will shift.

When expressed as a number, redshift is the difference in wavelengths divided by the original wavelength. So we see light from an object with a redshift of 1 with twice the wavelength it had when it was emitted; and a very distant object with a redshift of 9 has had its light stretched by a factor of 10.

Light can also be redshifted by climbing against gravity, or when an object is moving away from us. The latter is known as a Doppler shift, and can work the other way: an object heading our way, such as the Andromeda galaxy, can appear blueshifted.

10
Flashes and crashes

From across the cosmos we see flashes that mark the death of giants and the clash of dark stars. Some of these are among the brightest and most distant events visible, and give us insights into the nature of the universe.

The dark is rising

Some mysterious agent is pushing the universe apart. We don't know what it is. It is everywhere and we can't see it. It makes up more than two-thirds of energy in the universe, but we have no idea where it comes from or what it is made of. At least we have an evocative name for the stuff: dark energy.

Until two decades ago, we expected the expansion of the universe to be slowing down. But in 1998, astronomers got a shock when they analysed the brightness of distant type 1a supernovae (see Chapter 6). Because these explosions have a roughly known intrinsic brightness, they can be used as standard candles: the dimmer one looks, the farther away it is. And the unexpected dimness of many of these explosions implied that they were further away than was expected. Space seems at some point to have begun expanding faster, as if driven outwards by a repulsive force acting against the attractive gravity of matter (see Figure 10.1).

The other thing we can say for sure is that dark energy makes excellent fuel for the creative minds of physicists. They see it in hundreds of different and fantastical forms. The tamest of these is the cosmological constant, and even that is a wild thing. It is an energy density inherent to space, which within Einstein's general theory of relativity creates a repulsive gravity. As space expands there is more and more of the stuff, making its repulsion stronger relative to the fading gravitational attraction of the increasingly scattered matter. To account for the acceleration, we'd need a cosmological constant of about 1 joule per cubic kilometre of space. That's twice as much as all the dark matter plus normal matter put together.

Particle physics seems to provide an explanation for why empty space should have its own energy, through virtual particles that appear and disappear in the bubbling, uncertain quantum vacuum. The trouble is theory says these particles have far too high an energy density – in the simplest calculation, about 10^{120} joules per cubic kilometre.

This catastrophic discrepancy leaves the gate open for a herd of alternative theories. Dark energy could be quintessence: a hypothetical energy field that permeates space, changing over time and perhaps even clumping in different places. Or it might be a modified form of gravity that repels at long range. There are plenty of more arcane suggestions, such as that dark energy is caused by a loss of quantum information, or radio waves trillions of times larger than the observable universe.

It would help to know whether dark energy is changing over time. If it is, that would at least exclude the cosmological constant. In most models of quintessence, the energy becomes slowly diluted as space stretches; although in some it actually intensifies, pumped up by the universe's expansion. In most modified theories of gravity, dark energy's density is also variable. It can even go up for a while and then down, or vice versa.

The fate of the universe hangs in this balance. If dark energy remains steady, most of the cosmos will accelerate off into the distance, leaving us in a small island universe forever cut off from the rest of the cosmos. If it intensifies, it might eventually shred all matter in a big rip, or even make the fabric of space unstable here and now. Our best estimate so far, based mainly on supernova observations, is that dark energy's density is fairly constant. There is a suggestion that it is increasing slightly, but the uncertainties are too large for us to worry about this increase just yet.

Astronomers are now making a huge effort to pin down the behaviour of dark energy more precisely. Several projects, such as the Dark Energy Survey, are looking for several telltale signs of dark energy over a wide swathe of the sky. They are catching many more supernovae, plotting the positions of millions of galaxies to reveal fossilized sound waves from the big bang, and counting galaxy clusters through cosmic time, all of which should be affected by dark energy in different ways.

An even more impressive posse of dark energy hunters will set out in a few years, including dedicated space missions such as Euclid, and a new generation of giant telescopes: the Large Synoptic Survey Telescope, the Thirty Meter Telescope, the European Extremely Large Telescope and the Giant Magellan Telescope.

Few imagine that the hunt will be over soon. After two decades of puzzlement we still have no clue as to dark energy's identity. But on the bright side, we do have some clues to where the clues may lie.

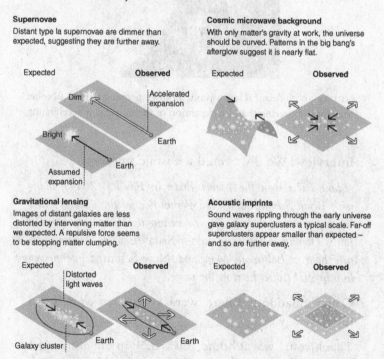

Supernovae

Distant type Ia supernovae are dimmer than expected, suggesting they are further away.

Cosmic microwave background

With only matter's gravity at work, the universe should be curved. Patterns in the big bang's afterglow suggest it is nearly flat.

Gravitational lensing

Images of distant galaxies are less distorted by intervening matter than we expected. A repulsive force seems to be stopping matter clumping.

Acoustic imprints

Sound waves rippling through the early universe gave galaxy superclusters a typical scale. Far-off superclusters appear smaller than expected – and so are further away.

FIGURE 10.1 Several lines of evidence suggests that something is accelerating the universe's expansion.

FIGURE 10.2 Adam Riess was one of three people awarded a Nobel Prize for discovering that the expansion of the universe is accelerating.

Interview: We discovered a cosmic mystery

Adam Riess won the Nobel Prize in Physics (2011), along with Brian Schmidt and Saul Perlmutter, for the discovery that the expansion of the universe is speeding up. Riess now works at Johns Hopkins University and the Space Telescope Science Institute, both in Baltimore, Maryland. New Scientist *interviewed him in 2011 after he won the prize.*

Congratulations. Where were you when you heard the news?

Thank you. I was at home. It was 5.30 in the morning. I was trying to sleep. My son is 10 months old and doesn't sleep so well. I was hoping he would fall back to sleep when I got the call. It was pandemonium at that point.

What was it that you, Schmidt and Perlmutter found?

We were two teams of astronomers who observed nearby and distant supernovae and used them to infer the amount of expansion of the universe at different times in its history. We determined that the universe, contrary to expectations, was not slowing down in its expansion — it was actually speeding up.

You were at the University of California, Berkeley, when you did the work. Perlmutter's team was in Berkeley too. Was there competition?

It was pretty competitive. We both knew that we were collecting the same kind of data and that it was the first of its kind. Neither of us wanted to be second or be so far behind that we were not involved. I would see those guys from time to time. We even socialized a little bit. Saul was kind enough to arrange a parking space for me at the Lawrence Berkeley National Laboratory [where he still works]. I would walk down the hill to go to work at the [University of California] Berkeley campus.

Einstein had the idea that space time has an intrinsic energy density that does not change with time, called the cosmological constant — but later called this notion his 'biggest blunder'. Is your work a vindication for him?

It's an impressive success of Einstein's general relativity. All these decades later, when we see very exotic phenomena in the universe, they can be fully accommodated, even expected in his theory.

It wasn't that far back that astronomy and astronomical observations were not even considered for the Nobel prize.

Right. I could certainly name discoveries from the past from cosmology that are completely Nobel-worthy: detecting the expansion of the universe or the scale of the universe, and the observations that indicate the presence of dark matter or some kind of additional gravity. These are fundamental to our understanding of physics.

You are only 41. What's next?

Well, there's still the literature prize this week, and the economics prize. I'm just kidding [laughs]. Before hearing about the Nobel prize, I had a couple of interesting projects in the works, and I'm going to keep working on them. They are things involving the Hubble Space Telescope and how we measure distances at a closer range.

Radio stars

Fast radio bursts (FRBs) are some of the universe's most elusive phenomena: powerful radio signals that flash from distant space for milliseconds and then disappear without a trace. They have been blamed on everything from black holes to extraterrestrial intelligence.

Because they're so brief, and because radio telescopes can only watch a small area of the sky at a time, only 18 FRBs have ever been detected. Of those, only one has been observed to repeat: FRB 121102.

In 2017 a team of astronomers finally pinpointed this repeating burst. Shami Chatterjee at Cornell University in Ithaca, New York and his colleagues tracked down the FRB using the Karl G. Jansky Very Large Array, a group of 27 radio telescopes in New Mexico, and the 21-telescope European VLBI Network. Together, these networks can achieve much higher resolution than any single radio dish, and the team located the FRB about 100,000 times more precisely than previous attempts.

This allowed them to clearly link the FRB to a dim dwarf galaxy around a tenth the diameter of the Milky Way, more than 2.5 billion light years away.

Knowing where an FRB comes from allows us to rule out some of the many proposed explanations for their origins. Since this example is so far away, it must be extremely energetic and bright. This implies that it's unlikely that any of the other FRBs we've seen come from our immediate neighbourhood, as some had proposed – although it is possible that FRB 121102 is special and that most FRBs are of an entirely different, non-repeating type.

Some exotic options for the origin of FRBs include exploding microscopic black holes, and clumps of dark matter colliding with black holes. A slightly more prosaic explanation for FRB 121102's repeated outbursts is that they come from an active galactic nucleus. But Chatterjee's preferred explanation is that FRB 121102 and its constant radio companion are caused by the remnants of a supernova being energized by a young, rapidly spinning neutron star. Since the FRB's host galaxy is similar to the surprisingly faint galaxies that produce the brightest supernovae, this scenario is an enticing fit – although it's far from proven.

Black hole birth cry

Gamma-ray bursts, which were first discovered in the 1970s, appear to come from random spots in the sky about once per day. They can release more energy in a few seconds than the sun will in its expected 10-billion-year lifespan.

The shortest bursts typically last less than a second and are now known to be caused by the merger of two neutron stars. Long bursts last for seconds to minutes and are thought to occur when massive stars explode as their cores collapse. They have been seen to coincide with very luminous supernova explosions. One burst in 2008, GRB 080319B, could just about be seen with the naked eye, despite being 7.5 billion light years away.

The theory is that in long bursts, the explosion fires out jets of matter at near light speed, emitting copious radiation in the process. When the jets happen to be pointed at Earth they send a narrow beam of radiation our way. Much of this is doppler shifted up to gamma-ray frequencies by the extreme speed of the jet.

What happens at the heart of the burst is still unclear. If the star collapses to form a rapidly spinning neutron star with a powerful magnetic field, called a magnetar, it could violently stir up matter around it to produce jets. Alternatively, if the star collapses to form a black hole, the jets might arise from the interaction of the black hole with matter spiralling into it. Either way, the tremendous rotational energy of the black hole or neutron star is thought to power the jets.

One way to distinguish between the two possibilities is to measure the total energy of the blast. The energy of a rotating object depends on its mass, and there is a limit to how heavy neutron stars can get before they collapse to black holes. There is no limit on the mass of black holes, however, so they could provide more energy than neutron stars.

In 2010, Bradley Cenko of NASA's Goddard Space Flight Center and colleagues analysed four of the brightest gamma-ray bursts detected by NASA's Fermi Gamma-ray Space Telescope.

The most powerful of the bursts, called GRB 090926A, released about 1.4×10^{45} joules of energy in its jets. Neutron stars should be able to produce no more than 3×10^{45} joules in total with only a small fraction of their energy going into jets. So the researchers argue that this burst, along with the three others, must be due to black holes instead. However, Stan Woosley of the University of California, Santa Cruz, points out that this may only apply to the most powerful gamma-ray bursts, while the creation of magnetars could power weaker ones.

Because gamma-ray bursts can be seen from so far away, some astronomers want to use them to probe the expansion history of the universe back to its early days, and perhaps help us find out how dark energy is behaving. Before that is possible we will have to understand them much better, so that we can work out the intrinsic luminosity of any burst by analysing the way its radiation fluctuates.

If we were desperately unlucky and a burst went off in our own galaxy and pointed at Earth, such power could wreak havoc. Gamma-ray bursts have been tentatively linked to mass extinctions. On the other hand they may have catalysed genetic mutations that helped life to diversify.

Spacequakes

Almost all of our knowledge of the heavens has come through electromagnetic radiation. Millennia of observing with visible light has been followed over the past century with instruments that detect radio, infrared and ultraviolet, X-rays and gamma rays. Neutrinos were the one exception – they reassure us that the reactor at the centre of the sun is still switched on, and soon they may be used to probe the hearts of active galaxies.

Then in 2015 a very different kind of signal was finally picked up, when the LIGO experiment saw the collision of two black holes about 1.3 billion light years away by picking up gravitational waves – the stretching and squeezing of space-time. It earned a Nobel prize in 2017 for three leaders of the collaboration, Rainer Weiss, Barry Barish and Kip Thorne.

The signal was picked up by LIGO's two observatories in Hanford, Washington, and Livingston, Louisiana, on 14 September 2015. The details of the waveform show how two black holes, around 36 and 29 times the mass of the sun, circled each other closer and closer until they finally merged into one.

The announcement caused a sensation among physicists and astronomers across the world. Gravitational waves will allow us to explore fundamental physics, examine the weirdest objects in the universe and possibly even peer back to the universe's earliest moments.

Since then, LIGO has picked up waves from more pairs of colliding black holes, one of which was also spotted by the Virgo gravitational wave detector, near Pisa in Italy. Having that second detector gives astronomers a much more precise direction for the source, which in future could help them track down any light or other radiation.

The second pair of merging black holes were about 8 and 14 times as massive as the sun – in the range that astrophysicists expected to exist as the product of collapsing stellar cores. But the others have shown us that there is a population around 25 to 35 solar masses, a group that we knew nothing about before the LIGO experiment began.

FIGURE 10.3 One hundred years after gravitational waves were proposed, LIGO detected them emanating from two black holes spiralling towards one another.

Then in August 2017, LIGO saw gravitational waves from colliding neutron stars for the first time. This event, 130 million light years away, was spotted by other astronomers. About 70 telescopes and observatories across the planet and in space turned in concert to face the same spot in the constellation Hydra, detecting a gamma-ray burst and a visible afterglow.

This was proof that neutron star mergers can cause short-duration gamma ray bursts. Astronomers also got their first sighting of heavy elements being formed, as about an Earth-mass of gold was thrown out of the explosion, along with other elements including uranium, plutonium and lead.

As more signals come in we should be able to get new insights into the history and composition of the universe as a whole. Several black hole mergers can be combined to help understand the nature of dark energy. From the shape of the signal – how the waves' frequency and volume rise and fall – we can discern the sizes of the black holes involved, and determine how loud the event was at its source. Comparing how powerful it really was to the faint vibrations LIGO detected tells us how far away it occurred. Combined with observations from standard telescopes, this can tell us how space has expanded during the time the waves took to reach us, providing a measure of dark energy's effect on space.

Other types of detector could come on the scene. The European Space Agency is planning the Evolved Laser Interferometer Antenna (eLISA), a huge, space-based detector that could pick up much longer wavelengths, including the waves that should be emitted when two supermassive black holes collide. Its technology has already been tested in a preparatory mission, LISA Pathfinder.

Further ahead we might see detectors working at shorter wavelengths than LIGO, which may allow us to sense primordial gravitational waves from the very young universe. These waves should have been produced in the period of inflation – the tremendous growth spurt in the first instants after the big bang. Such observations may even point the way toward a grand unified theory of the universe. Once, the four fundamental forces were united into a single force. As the universe expanded and cooled, the forces split off from one another in a series of as-yet poorly understood events: events that gravitational waves could probe.

What is the most distant object we have seen in the universe?

Perhaps surprisingly, the record holder (at time of going to press) is not an intense gamma-ray burst or an ultraluminous quasar, but a small galaxy. GN-z11 has a redshift of 11.09, comfortably greater than that of any other known object.

If you want to know what that means in terms of light years, the answer isn't quite so simple. One measure is light travel time: the light from GN-z11 has been travelling for 13.4 billion years. But since that light set out, the universe has been expanding, which changes things. Among several different measures of distance used by cosmologists, probably the nearest to an intuitive notion is called proper distance. By this measure, GN-z11 is about 32 billion light years away.

Conclusion

Even in a spacecraft that travels at almost the speed of light, there is a limit to how far we can get. The universe is expanding and accelerating, dragging distant objects away from us at an increasing rate. The furthest galaxies, quasars and gamma-ray bursts we see today are already out of reach. Features marked out in the cosmic microwave background, afterglow of the nascent universe, have today become clusters and superclusters of galaxies, but they are even further beyond us.

So after our tour we will have to head home and explore through time instead: into a distant future where the sun dies, the Milky Way and Andromeda collide, and most of the observable universe is dragged beyond the cosmological horizon to invisibility. If we wait long enough, we might even find out what dark energy is.

Fifty ideas

This section helps you to explore the subject in greater depth, with more than just the usual reading list.

Four things to do and see

1 Go into darkness. City lights and pollution wipe out the night sky for most of us, but you don't have to go far to find a better place. Check http://www.darksky.org/idsp/reserves/ for official dark sky reserves, parks and communities, such as Rhön in central Germany, and Sark, the world's first dark-sky island. Kielder Forest, in Northumberland, UK, even has its own public observatory: https://www.kielderobservatory.org/. Or just get as far as you can from gaudy civilization.

2 If you are in a dark enough place to see the Milky Way, lie down and convince yourself that it is just the accumulated light of millions of stars more distant than the ones we see clearly from Earth. Then look for two planets, or the moon and a planet, to trace out the plane of the Solar System. Finally, get vertigo.

3 Look out for a meteor shower. The Perseids in August are among the strongest and longest showers, lasting for many days. Each streak of light is a fragment of the comet Swift–Tuttle, heating up as it enters Earth's upper atmosphere at 58 kilometres per second.

4 If you'd love to walk on Mars but don't have the billions needed to cobble together a mission, then Devon Island in Canada may be for you. Cold, dry and cratered, it has an almost Martian climate and landscape, and NASA tests Mars-mission technology there. For an alternative pseudoMars, try the highlands of Iceland.

Ten historical sites and modern observatories worth a visit

1 The Copernicus Museum in Torun, Poland. Discover plenty about the astronomer who laid the foundations for the scientific revolution by suggesting that the Earth goes around the Sun: www.visittorun.pl/301,l2.html

2 The island of Ven, located between Sweden and Denmark, holds the reconstructed observatory of Tycho Brahe, whose detailed observations gave Johannes Kepler the data to derive his laws of planetary motion: www.tychobrahe.com/en/

3 Woolsthorpe Manor in Lincolnshire, UK, was home to another hero of the scientific revolution. Sir Isaac Newton took refuge in the family home when the plague struck Cambridge in 1665, working here on the ideas that would change the face of physics and mathematics: www.nationaltrust.org.uk/woolsthorpe-manor

4 The Royal Observatory, Greenwich, London is where the crosshairs of George Airy's telescope defined Earth's prime meridian, the line of zero degrees longitude. The astronomy centre and Great Equatorial Telescope are two more of the observatory's treasures: www.rmg.co.uk/royal-observatory

5 Ulugh Beg's observatory in Samarkand, Uzbekistan. Built in the 1420s, the site housed three enormous astronomical instruments that were left to rot until excavated in 1908. Now it's a UNESCO World Heritage Site: www3.astronomicalheritage.net/index.php/show-entity?idunescowhc=603

6   A panoply of instruments on the summit of Mauna Kea, Hawaii, which includes the twin Keck telescopes with their 10-metre mirrors. Gen up and go star gazing at the Mauna Kea Visitor Information Station: www.ifa.hawaii.edu/info/vis/visiting-mauna-kea/visitor-information-station.html

7   Another cluster of impressive telescopes perches on Roque de los Muchachos, the splendidly named high point of La Palma in the Canaries: www.visitlapalma.es/en/recursos_culturales/observatorio-astrofisico-roque-de-los-muchachos/

8   The Lovell Telescope at Jodrell Bank Observatory in Cheshire, UK, is a veteran of more than 60 years, but remains the third largest steerable radio dish in the world – after the Green Bank Telescope in the US and Effelsberg in Germany. Start at the Discovery Centre: www.jodrellbank.net/

9   ALMA, an array of dishes that form the most sensitive detector of submillimetre waves, is 5000 metres high in the dry Atacama desert of Chile. For slightly less of a headache risk, the telescopes of La Silla are at only 2400 metres: www.almaobservatory.org/en/outreach/alma-observatory-public-visits/

10 One to plan for is the Extremely Large Telescope, also in the Atacama. Now under construction, by 2024 it will be the biggest of a new generation of giant optical telescopes. Its primary mirror will be nearly 40 metres across: www.eso.org/public/teles-instr/elt/

Five astronomical numbers

1 About 1.3 million Earths would fit into the volume of the sun.

2 There are hundreds of billions of stars in our galaxy, and hundreds of billions of galaxies in the observable universe.

3 The brightest gamma-ray bursts can briefly appear 1000,000,000,000,000,000,000 (one thousand billion billion) times as luminous as our sun, although four of those zeros are because their emission is concentrated into a narrow beam.

4 The pulses from some pulsars are so reliable that they could be used to keep time to an accuracy of better than one microsecond per decade.

5 The solar system orbits the centre of our galaxy in roughly 250 million years, a timespan sometimes called the galactic or cosmic year. By this reckoning, Earth is about 18 years old.

Sixteen deep space quotations

1 'The sun
   Exalted on his throne bids all things tend
   Toward him by inclination and descent,
   Nor suffer that the courses of the stars
   Be straight, as through the boundless void they move,
   But with himself as centre speeds them on
   In motionless ellipses.'

   Edmond Halley

2 'Several billion trillion tons of superhot exploding
   hydrogen nuclei rose slowly above the horizon and
   managed to look small, cold and slightly damp.'

   Douglas Adams, Life, the Universe and Everything

3 'Damn the Solar System. Bad light; planets too distant;
   pestered with comets; feeble contrivance; could make
   a better myself.'

   Francis Jeffrey, Scottish judge and critic

4 'Having probes in space was like having a cataract
   removed.'

   Hannes Alfvén, Swedish Nobel laureate in physics

5 'Space isn't remote at all. It's only an hour's drive away if your car could go straight upwards.'

Fred Hoyle, British astronomer

6 'There is no easy way from the Earth to the stars.'

Seneca the Younger, Roman philosopher,
dramatist and humorist

7 '...a bullet would spend almost seven hundred thousand years in its journey between us and the fixed stars. And yet when in a clear night we look upon them, we cannot think them above some few miles over our heads.'

Christiaan Huygens, Dutch mathematician and scientist

8 '...we make guilty of our disasters the sun, the moon, and the stars, as if we were villains by necessity, fools by heavenly compulsion, knaves, thieves, and teachers by spherical predominance, drunkards, liars, and adulterers by an enforced obedience of planetary influence...'

William Shakespeare

9 'The stars are majestic laboratories, gigantic crucibles, such as no chemist could dream.'

Henri Poincaré, French mathematician,
scientist and philosopher

10 'One important object of this original spectroscopic investigation of the light of the stars and other celestial bodies, namely to discover whether the same chemical elements as those of our earth are present throughout the universe, was most satisfactorily settled in the affirmative; a common chemistry, it was shown, exists throughout the universe.'

William Huggins, English astronomer

11 'Whereas all humans have approximately the same life expectancy, the life expectancy of stars varies as much as from that of a butterfly to that of an elephant.'

George Gamow, Ukranian theoretical physicist and cosmologist

12 'Organs blaring and fugues galore, Kepler's music reads nature's score'

A mnemonic for the spectral sequence of stars in Chapter 5.

13 'On bad afternoons, fermented grapes keep Mrs Richard Nixon smiling'

Another.

14 'We find them smaller and fainter, in constantly increasing numbers, and we know that we are reaching out into space, further and ever further, until, with the faintest nebulae that can be detected with the greatest telescope, we arrive at the frontiers of the known universe.'

Edwin Hubble, American astronomer

15 'The effort to understand the universe is one of the very few things that lifts human life a little above the level of farce, and gives it some of the grace of tragedy.'

Steven Weinberg, Nobel laureate in physics

16 'Space is big. Really big. You just won't believe how vastly, hugely, mind-bogglingly big it is. I mean, you may think it's a long way down the road to the chemist, but that's just peanuts to space.'

Douglas Adams

Five jokes, as tenuous as interstellar gas

1 Why does a carbonaceous chondrite taste better than a chunk of limestone? It's a little meteor.

2 Why isn't the Dog Star laughing? It's Sirius.

3 How does the man in the moon cut his hair? Eclipse it.

4 I was up all night wondering where the sun had gone and then it dawned on me.

5 Why is there no Nobel prize for astronomy? It would only be a constellation prize.

Five stories of the star-gazers

1 Meticulous astronomer Tycho Brahe lost his nose in a duel, and had it replaced with a metal one. The prosthetic was once thought to be silver or gold, but a 2012 analysis of the body reported that it was actually made of brass.

2 Fritz Zwicky, the first person to find evidence for dark matter, once got his assistant to shoot a rifle out of the Palomar observatory to see whether he could track the bullets with the Hale Telescope.

3 In 1962, two French astronomers were excited to discover that a star suddenly began to emit bright spectral lines associated with the element potassium. They knew that other stars did the same. Was this a new class of potassium flare star? Eventually the flares proved to be much closer to home: matches struck to light the cigarettes of observatory staff.

4 Lyman Spitzer, who developed the concept of space-based telescopes and is commemorated in the name of NASA's infrared Spitzer Space Telescope, also reached for the skies as a mountaineer. On Baffin Island in northern Canada, he made the first ascent of Mount Thor by its north ridge. And to the consternation of the authorities at Princeton University, New Jersey, in 1976 at the age of 64 he conquered Cleveland Tower, the tallest building on campus.

5 Galileo's famous line '*eppur si muove*' ('and yet it moves') was supposedly delivered at his 1633 trial for heresy in Rome, after his enforced recantation of the Copernican

model of the solar system. There is no evidence he actually said it… but perhaps he has been granted a posthumous act of defiance instead. His severed middle finger is preserved in the Galileo museum in Florence, in an upright attitude that might be interpreted as less than polite.

Five online resources

1 An excellent blog on all things astronomical: www.armaghplanet.com/blog/

2 NASA's compendious website covering exploration and astronomy: www.nasa.gov/

3 Maps of the sky: www.skymaponline.net/

4 Watch our solar system in action: https://theskylive.com/3dsolarsystem

5 Contribute to astronomy by helping to classify galaxies: www.galaxyzoo.org/

Glossary

Accretion The process of accumulating matter, usually by gravity.

Active galactic nucleus The luminous central object in many galaxies, powered by the accretion of matter by a super-massive black hole. There are several classes of active galactic nucleus, including seyferts, quasars and radio galaxies.

Astronomical unit (AU) The average distance between Earth and sun, about 150 million kilometres.

Black hole The result of total gravitational collapse, according to the general theory of relativity: a region of spacetime cut off from our vision by a boundary called the event horizon, from which nothing can escape.

Coma The roughly spherical bright area of dust and gas blown from a comet's nucleus.

Corona The outer part of the sun's atmosphere, at a temperature of more than a million kelvin.

Cosmic rays High-energy charged particles originating outside the solar system.

Dark matter Invisible material that is thought to account for the motions of galaxies and clusters, outweighing ordinary matter in the universe by about five to one.

Dwarf planet An object orbiting the sun that has enough mass for its own gravity to force it into a nearly round shape, but has not cleared other material from the neighbourhood around its orbit (and is not a moon).

Eclipse An event in which one body blocks some or all of the light from another (as in a solar eclipse or planetary transit), or casts a shadow across it (as in a lunar eclipse, when the Earth's shadow falls across the moon).

Elliptical galaxy A galaxy with a rounded shape, without spiral arms, where there is little new star formation.

Globular cluster A spherical group of many thousands of stars that orbits the Milky Way or another galaxy.

Jet A narrow stream of matter moving at high speed, sometimes close to the speed of light, often fired from a black hole or other dense object.

Kuiper belt The ring of icy objects beyond Neptune, including Pluto.

Local group A loose association of galaxies including the Milky Way, Andromeda, Triangulum and a few dozen dwarf galaxies, about 10 million light years across.

Luminosity The total power output of an object.

Magellanic Clouds Two irregular dwarf galaxies orbiting the Milky Way.

Mare From the Latin for sea, this term is used for the huge dark impact basins that dominate the near side of the moon, and for actual seas on Saturn's moon Titan.

Metal In astronomy, all elements except hydrogen and helium.

Molecular cloud An interstellar cloud that is cool and dense enough for hydrogen to form molecules of H2; they are often the sites of star formation.

Neutron star The ultra-dense remnant of some types of supernova explosion, made up mainly of neutrons.

Oort cloud A vast halo of icy objects suspected to surround the solar system, the source of long-period comets.

Planetary nebula A glowing shell of gas that has been ejected in the late life of a giant star, often illuminated by a remaining white dwarf within.

Plasma High temperature ionised gas. Fully ionised plasma is composed of free electrons and free atomic nuclei. It forms the bulk of most ordinary stars, and much of the tenuous interstellar medium.

Quasar A very luminous class of active galactic nucleus.

Red giant A stage in the life of a star when it stops fusing hydrogen in its core. The core contracts while the outer layers of the star expand and cool.

Redshift A shift in the wavelength of radiation towards the red end of the spectrum. May be caused by the source moving away from the observer, by the expansion of space, or when light travels up through a gravitational field.

Stellar wind Plasma blown out from the atmosphere of a star.

Supernova remnant The gas blown out by a supernova explosion, which can form a glowing nebula for thousands of years.

Supermassive black hole A black hole millions to billions of times the mass of the sun. Such black holes are found in the cores of almost all large galaxies.

Transit The passing of one object across the face of another.

Trans-Neptunian object Anything orbiting the sun beyond Neptune.

White Dwarf The dense remnant of a medium-sized star that is no longer supported by nuclear fusion.

Picture credits

Figure 1.1 NASA/SDO

Figure 2.1 U.S. Geological Survey/ Arizona State University/ Carnegie Institute of Washington/ Johns Hopkins University Applied Physics Laboratory/ NASA

Figure 2.2 NASA/JPL/University of Arizona

Figure 3.1 NASA/SWRI/MSSS/Gerald Eichstädt/Seán Doran

Figure 3.2 NASA/JPL/Space Science Institute

Figure 3.3 NASA/JPL/Space Science Institute

Figure 3.4 NASA/ESA. Acknowledgement: T Cornet, ESA

Figure 4.1 NASA/JHUAPL/SWRI

Figure 4.2 NASA/JHUAPL/SWRI

Figure 5.2 NASA/ESA/Hubble and the Hubble Heritage Team

Figure 6.1 Daily Herald Archive/SSPL/Getty

Figure 6.2 NASA, ESA, NRAO/AUI/NSF and G Dubner (University of Buenos Aires)

Figure 7.1 NASA/JPL-Caltech/R Hurt

Figure 7.3 NASA/ Goddard Space Flight Center

Figure 8.1 Pikaia Images

Figure 9.1 X-ray: NASA/CXC/UAH/M. Sun et al; Optical: NASA, ESA, & the Hubble Heritage Team (STScI/AURA)

Figure 10.2 Michael Reynolds/Epa/REX/Shutterstock

Figure 10.3 R Hurt/Caltech-JPL

Index